# Shadows From Can Tho

**Shadows From Can Tho**



ISBN: 978-0-9830654-1-8

Printed in the United States of America by:

The Print Shop, Medina, New York

# *Shadows From Can Tho*

Stories of War and Love
in Vietnam

*J. R. Barth*

## Dedication

To all those who understand that the ultimate Truth of the Universe is goodness, and who are not afraid to pursue it, this book is dedicated.

*Though the cause of evil prosper,*
*Yet 'tis Truth alone is strong;*
*Though her portion be the scaffold,*
*And upon the throne be wrong,*
*Yet that scaffold sways the future,*
*And behind the dim unknown*
*Standeth God (Good) within the shadow*
*Keeping watch above His own.*

Adapted from "The Present Crisis"
written by James Russell Lowell in 1848

## Acknowledgements

The motivation for this work belongs to my children: Richard C.R. Barth, Sarah L.R.B. Dressel, and Elizabeth J.R. Barth; my stepsons, Geoffrey M. and David N. Bowman; and especially to my wife, Jessie McNall Barth. Without their encouragement, patience and love, this memoir would not have happened.

People who helped me with details include my former wife, Jody, who provided more than 120 letters I wrote to her from Vietnam in 1970 and 1971 that were helpful in recalling specific dates, people and locations. Also Tom Wilkes, who provided his recollection of the accident that took the life of my flight school classmate, Jack Bagley, in October 1970.

And many, many thanks to my editors, who wished to remain anonymous. You know who you are!

## Character Names

All the stories in this memoir are real and written as the author recalled them. The names of most of the characters have been changed to respect their privacy, and those that were not changed have been noted with a *.

## Cover Credits

The cover was designed by the author with technical assistance from Alec Dressel. The orange background is a photograph of a silk scarf purchased in Can Tho, Vietnam in March 2013 by the author.

## Table of Contents

# Introduction

This book contains dozens of stories about my first-hand experiences as a helicopter pilot in the 191st Assault Helicopter Company in Can Tho, Vietnam from May 1970 to May 1971.

The stories are about:

- What it was like to fly a sophisticated state-of-the-art machine in combat;
- What it felt like to be shot at and shot down, and to walk away from it;
- The characters and personalities of the men I flew with and lived with everyday, for a year;
- The Vietnamese people and how I went from hating to loving them;
- How warfare permanently altered my perspective and changed my values; and
- My personal successes and failures, frankly told.

Many readers who are not familiar with combat may get the impression that the stories in this book describe the worst experiences a soldier can have in war. But my experiences are pale in comparison to what many soldiers and pilots endured during their tours of duty, and I owe them my respect.

The original reason for this writing was to satisfy my family's curiosity about the Vietnam War and what my role in it was, hoping it would help my kids understand what makes the old man tick. But the reason for writing it became more complicated when I realized, a year or so later, that the United States created a terrible legacy in Southeast Asia when it started a war there, abandoned the South Vietnamese to defend themselves, and didn't clean up the mess it created when it left. That mess exists today in the form of unexploded ordnance (UXO) and landmines that Americans left behind, a mess that still kills and maims dozens of Vietnamese, Cambodians and Laotians every year. This is our legacy; it's the shadow that falls across Southeast Asia almost half a century later.

The book title was chosen for two reasons. First, because these stories exist as memories of a time long past that persist in connecting me with that time and that place. They are my shadows, my permanent Vietnam connection. The second reason has to do with the old adage, what goes around, comes around; or to say it another way, actions have consequences. In my view the U.S. legacy in Vietnam has spread eastward across the Pacific, and we Americans are reaping what we sowed there: ongoing terror and government intrusion to an unprecedented degree. The shadow that we cast in Can Tho and throughout all of Southeast Asia has come around, and now we too stand in its darkness.

We created this problem; we can fix it. To find out how, please refer to the Postscript at the end of the book.

A portion of the proceeds from the sale of this book are used to help finance the removal of unexploded ordnance and landmines in Southeast Asia.

J. R. Barth
June, 2013

Notes:

There are song titles surrounded by boxes in the text that refer the reader to pop music that was at or near the top of the music charts in 1970-1971. The songs have lyrics that relate to the story material and the music adds appropriate and interesting flavor.

The omega sign (--- Ω ---) throughout the text simply separates the end of one story, or part of a story, from the beginning of another.

## Chapter 1 - MAY 1970

On the Continental Airlines flight that I boarded at Travis Air Force Base it was too hot, and then too cold. The food was good, and then it wasn't. I think it was steak. A couple of flight school classmates who were on the same flight had saved a seat for me near them, so I was among friends.

An infantry lieutenant seated near me who I didn't know couldn't stop shaking and wanted to talk out his nervousness. But our conversation didn't seem to help much. I had a guilty sort of sympathy for him, flush with gratitude that I was going to fly choppers and not lead troops on foot through the jungles and rice paddies of Vietnam like he was. But I too was nervous, and at times claustrophobic. The anticipation was awful and I just wanted to get out of that plane as soon as I could.

There was a thirty minute stop in Honolulu for fuel and a crew change, then the flight resumed. The new flight attendants were a mean, unfriendly bunch. I managed several hours of fitful sleep before the pilot announced he was going to land at Clark Air Force Base in the Philippines to refuel. After landing we deplaned to stretch and I remember vividly how hot and humid the air was, even early in the morning.

As I walked down the plane's steps the sky was mostly clear and the sun was just coming up behind a gorgeous, black mountain to the east, probably Mt. Arayat. Blocked by the immensity of the mountain, the sun threw its bronze-colored rays north and south while a few soft, gray clouds hanging near the peak added a dewy quality to the sunrise. The beauty of that morning seemed to be entirely out of place, and I wondered how something so beautiful might exist so close to the battlefield. (Later I found beauty of another sort much nearer the battlefield.)

After an hour we reboarded the Continental jet and took off for Saigon, touching down at 9:00 AM on Monday, 4 May, 1970 at Bien Hoa (*bin wah*) Air Force Base. The last leg of the flight had been extremely quiet with only an occasional question from a flight attendant breaking the silence. Somewhere between Travis and Bien Hoa, a mental and emotional threshold had been crossed. I was no longer afraid, but I was anxious to get settled in a unit to do the job I'd been trained for. I was tired of traveling and waiting, and waiting and traveling. I just wanted to get on with whatever was in store.

Leaving the plane in broad daylight we found the environment to be surprisingly normal. There were all sorts of planes landing and taking off, and vehicles of all kinds running back and forth. The wide cement runways were joined with numerous taxiways, and beyond them were countless hangars and revetments (barricades between choppers to protect them from rocket or mortar attack) and buildings of all sizes and types. But some things were unusual: the sun was terrifically bright, so bright it pierced even a closed eyelid, and like Clark AFB the air was hot and humid. But by far the most unusual thing was the odor of the place which could only be described as a stench, unlike anything I'd smelled before. It seemed like a combination of rotten garbage, burned jet fuel and dung, and it was everywhere.

Dark green buses that looked like school buses, with heavy wire mesh covering the windows, carted us and our duffle bags seven miles to a large Quonset hut that was the office of the 90th Replacement Battalion at Long Binh Army Post. One of only two places where men fresh from the States reported for duty, Long Binh (*long bin*) was a gigantic U.S. Army base near Saigon. The 90th was the permanent duty point on the original orders I had received when I entered active duty more than a year before. It was the last known destination for me, and where the path led after that was a mystery.

After almost 24 hours of travel, I was tired and felt greasy all over. Unaccustomed to the heat and humidity, I, like all the other GIs there, was sweating like a condemned man and by now my khaki uniform was half soaked with perspiration. The window air conditioner in the office was cranked up all the way, but it couldn't keep up with all the traffic going through the place. The stench outside had made its way inside, but now it was mixed with body odor, and in that congested space it made me gag with every breath.

The line making its way to the front of the reporting desk moved slowly but persistently, with each man turning his orders over to the officer in charge (OIC) and getting a bunk assignment in Long Binh's temporary quarters. When I reached the desk, I was astounded to find that the OIC was Lt. Rusty Miller who had been a member of the Ohio State Glee Club at the same time I was. Rusty was a great guy, about 5'9" tall with brown hair and a big smile to go with his perpetual friendliness. I think we were equally surprised to see each other half way around the world, and I regretted we didn't have time to get reacquainted. (Our paths would cross again more than a decade later when Rusty called me at home by telephone, from Rome.)

The officers' temporary housing was just like the enlisted men's, except they were one-story instead of two. The buildings were long and wooden with slatted walls to let air through, and wire screen at the top above the slatted walls. Army cots were lined up on each side of the concrete floor from one end to the other and there was a screen door at each end of the building. The roof was plain, unpainted tin and there were sandbags piled four feet high around the outside walls. It was obviously designed to provide as much safety as possible while trying to offer some comfort as well, but to a guy like me from a northern climate, it was just plain hot, both day and night, and sleeping was almost impossible.

I found the shower and washed off the greasy feeling, then went to the supply office to get my basic issue of olive drab equipment: jungle fatigues, underwear, socks, etc., and afterward went back to my bunk and waited. I took an anti-malaria pill which made me sick, and spent the night sweating in my bunk.

That same day, 4 May, there was a tragic shooting on the Kent State University campus in Kent, Ohio where National Guardsmen opened fire on a group of students demonstrating against the war, and four students were killed. The tragedy seemed to be the culmination of growing unrest on college campuses across the country, and I remember thinking that it didn't matter whether I was in Vietnam or America, because Americans were now being killed as a result of the war in both places.

On Tuesday, 5 May, my friend from Ohio State and flight school, Tom Yost* arrived about 4:00 PM and we spent the next day waiting to get orders for our permanent duty stations. We went to the officers club and found out there was a swimming pool and sauna nearby, but neither of us were interested and all we really could do was wait-- and sweat. A couple of officers we knew from Officer Basic Training class and flight school passed through while we were there, but mostly we waited, and perspired, and wondered where we were off to next.

It was amazing that our journeys had essentially been side-by-side since the night we met in St. John Arena in March 1969. Now, more than a year later as we thought about our final destinations, both of us figured we would be split up and sent to different places and our side-by-side military experience would be over. But to our continuing amazement, we were both assigned to the 164th Aviation Group, headquartered at Can Tho (*cun tuh*) Airfield in the IV Corps operations area south of Saigon.

At 11:00 AM the next day we found ourselves on an Air Force transport plane en route to Binh Thuy Air Force Base with flight classmate Murray Majors, a group of enlisted men, and a lovely young lieutenant named Marti Macahilig, a petite Army nurse of Philippine descent with deep olive skin, coal black hair and beautiful dark eyes. We found out that she was from Oregon and was bound for hospital duty at Binh Thuy (*bin too-ey*) Air Force Base near Can Tho. But the conversation with Lt. Macahilig was stiff and awkward, no doubt reflecting the uncertainty and anxiety we all felt. In spite of being dressed in jungle fatigues and combat boots, she was very attractive, and as I summed up the situation someone Tom should be interested in!

**Binh Thuy - Can Tho - Soc Trang - Can Tho**

We touched down 45 minutes later at Binh Thuy AFB, deep in the middle of the Mekong (*may-kong*) Delta region of South Vietnam, and a jeep was there that took us to Can Tho Airfield and the 164th Headquarters. Unfortunately, we had to say goodbye to lovely Lt. Macahilig. The driver dropped us off in front of a well-maintained, one-story, white building with two tall flag poles, palm trees and banana plants out front, that looked like it could have been in the States somewhere. We walked up the concrete sidewalk and once inside were greeted by the OIC who took our orders and sent us to the mess hall a couple of blocks away. While there the Group S-1, a Major Ingleman, came in to talk with us to see if we had any desires about the type of duty we would be assigned. Murray wanted to fly gunships, but neither Tom nor I had any preference.

Ingleman left and suggested we might want to get a drink at the Can Tho Officers Club until he could finish our assignments, so we headed there for a beer. At the club we ordered a cold Budweiser, a welcome break from the heat, and Tom and Murray played bumper pool while I watched from the bar. A few minutes later the club door opened and in came the familiar face of 1LT Allen Hoppes, an amiable flight school classmate of ours from Ft. Wolters and Hunter Army Airfield! It seemed almost too good to be true! Not only had I been greeted by someone I knew when I reached the 90th in Saigon, I was being greeted by someone I already knew in Can Tho as well.

The four of us sat and talked about various potential assignments, then went to see Ingleman for our orders. Murray was heading for a cavalry outfit to fly gunships, and Tom and I were assigned to the 13th Combat Aviation Battalion headquartered in Soc Trang (*sock chrahng*).

About 2:00 the next morning we were flown by helicopter to Soc Trang where we reported to the 13th Battalion Headquarters and that's where the serendipity of side-by-side travel ended. Tom was assigned to the 121st Assault Helicopter Company (AHC) at Soc Trang Airfield, southeast of Can Tho about 30 miles, and I was sent back to Can Tho to the 191st AHC. While we expected our joint travel would end at some point, it was none-the-less difficult saying goodbye to Tom when I left Soc Trang. We really had no way of knowing if we would ever see each other again.

It was a short Huey ride back to Can Tho and a brief walk to the 191st headquarters where Al Hoppes was working as the company admin officer. After talking to Al for a few minutes I was sent in to meet the company commander, Major Able Sechrist, a humorless, sober, round faced officer with a pasty complexion and a pineapple brush cut, who occupied the chair behind his desk as if he were the Emperor of Can Tho. I stood at attention in front of his desk and saluted with the best military bearing possible as a sign of respect. Sechrist returned the salute and gave a very cold welcome to the 191st.

I was then introduced to the company Executive Officer, Captain Ted Everest, and then to the company's clerical staff. This was followed by a stop at the company supply office where I met the supply officer, 1LT Connie Leopard, and was issued additional gear: a poncho liner (which served as a bed blanket), a Smith and Wesson .38 revolver, a SPH-4 flight helmet, three Nomex flight suits, a pair of flight gloves, a flight calculator, aviation maps, etc. The one thing not available was a pair of boots in my size (13), so I had to use the boots I had brought from the States. I was then temporarily assigned to bunk with Captain Ron McConkey until a permanent bunking arrangement developed.

**First Duty**

The next morning, Sunday, 10 May, as I was writing a letter to my sister, Al interrupted and asked if I wanted to run an errand to Saigon for him. Since I was not on flight status yet and having nothing really important to do, I agreed to escort the remains of three 191st crew members to a Graves Registration Unit at Tan Son Nhut (*tahn sahn noot*) Air Force Base. They had been killed the previous day in a Huey crash near Tra Vinh, east of Can Tho. The remains were of Warrant Officer Terry Henry* from Clarion, PA (pilot), SP4 Kris Perdomo* (door gunner), and Henry's copilot (name unknown). The remains of SP4 Stephen Haight* (crew chief), also killed in the crash, were never found.

The formal orders for the trip were written and then signed by Maj. Sechrist. I signed a receipt for the body bags containing the remains as if they were just ordinary freight that had to be accounted for, and went to the flight line where I boarded a C-123 and sat with my "cargo" next to me. Each bag was about seven feet long and 3 feet wide, made of heavy black rubber with a single zipper that ran from top to bottom in the middle of one of the flat sides. The lumpiness of the bags made it clear something was in them. They didn't look like they contained complete bodies though, so I was curious as to what they actually did contain.

An hour later as the door of the C-123 opened at Tan Son Nhut, an ambulance met me. Two enlisted men loaded the bags into their hot ambulance through the rear doors and I crawled in the back with the bags. In a couple of minutes we were at the door to the Graves Registration building where they carried the bags in and carefully, respectfully, placed them on stainless steel tables.

A clerk was sitting at a desk in the corner. Forty tables ran up one side of the building and down the other in assembly line fashion, empty except those occupied by the remains of Henry, Perdomo and the copilot.

In contrast to the ambulance, it was incredibly cold in the concrete block building, I suppose to help preserve bodies. I gave my paperwork to the clerk who signed a form and gave me a copy as receipt for the "freight." Twenty feet away, a couple of other men were putting on rubber aprons and gloves, and then began to unzip the bags. I turned my head and glanced into the first bag, saw only a black flight helmet and a lot of muddy ashes, and that was all. No body, no flight suit, nothing. I avoided looking into the other two bags and sat down, waiting for the driver to come in and take me back to the flight line. As I waited I started to shake, either from the cold air or the impact of the situation, or maybe both. This was a new aspect of my Army experience, combat death and how it was handled. It was efficient, impersonal, cold, and very sobering.

I was dropped by jeep at the transport terminal to catch a ride back to Can Tho, but the clerk at the desk said "there ain't no more" flights scheduled to go to Can Tho for the rest of the day, and I should go to Hotel-3 (H-3) to catch a chopper ride. I asked him what H-3 was and he said the Tan Son Nhut heliport. So I took the bus to H-3 and waited for a flight, but by 7:00 PM there were none and the heliport office door had been locked for the night.

I walked to the officers club and had a couple of beers and tried to get into the dining room to have dinner, but it was full. I waited awhile and then left and went back to H-3 and slept overnight on a bench outside. The next morning I called Al and he suggested I go back to the Tan Son Nhut terminal and get orders made out for the next available Air Force flight to Can Tho. I arrived about 4:00 in the afternoon, tired, hot, hungry and depressed.

Going to Tan Son Nhut had been awful and awesome, and I couldn't stop thinking about the families of the casualties and how they would feel when they got the news. I was struggling with what seemed to be a gruesome reality, and hoped that the lyrics to Ray Stevens' current pop hit, *Everything is Beautiful*, were right, and that we were "gonna get it all together now; everything gonna work out fine."

| *Everything is Beautiful* - Ray Stevens |
|---|

**Boomerangs**

There were about 40 pilots in the 191st, 27 of whom were warrant officers or chief warrant officers and the rest commissioned officers. Of the commissioned variety there was one major, four captains, and eight lieutenants. There were also about 120 enlisted men of various ranks from private first class (PFC) to first sergeant (1SG). The officers lived in two-man rooms in single story wooden barracks, which were fortunately only a stone's throw from the officers club. The enlisted men lived in two-story barracks buildings, dormitory style, next to the 191st headquarters and near the mess hall, the supply office and the flight line.

All Assault Helicopter Companies (AHCs) were composed of 4 platoons of pilots, crews and support personnel, and about 35 Huey aircraft. Each company had a radio call sign like "Tigers" or "Vultures." The 191st call sign was "Boomerangs" and the company motto was "Boomerangs always come back." Based on my trip to Saigon with the three body bags, I realized the motto didn't guarantee in what condition that would be.

The primary mission of an AHC was to fly combat infantry troops from pickup zones (PZs) to landing zones (LZs) and back. Usually the first two platoons did the troop lifts with D or H model Hueys (referred to as "slicks") and the third platoon provided fire support with C model Hueys, a model that was specially designed for that task (referred to as "guns"). The fourth platoon provided maintenance services for the company aircraft. However, in the 191st the missions were slightly altered. The first, third and fourth platoons had the standard missions, but the second platoon's mission was to fly high ranking officers and guest VIPs throughout the IV Corps area. It was a cushy assignment for the pilots and crews who were expected to avoid actual combat conditions in order to protect the VIPs.

The radio call sign for the first platoon (slicks) was "Boomerang," the second platoon (VIPs) used "Green Delta," the third platoon (guns) was "Bounty Hunter," and the fourth platoon (maintenance) used "Wingnut." Each individual pilot had his own call sign number: the company commander was always Boomerang 6 and the executive officer, or second in command, was always Boomerang 5. The first platoon commander was always Boomerang 16 and other members of the platoon were 15, 14, 19, etc. The second platoon's commander was always Green Delta 26 and the various members of the platoon were Green Delta 25, or 24, etc. The third platoon's commander was always Bounty Hunter 36 and the other gun pilots were Bounty Hunter 35, or 34, etc. Out of step with the usual numbering, the fourth platoon commander was Wingnut 6 and the other maintenance pilots were Wingnut 5, or 4, etc.

**Keeping the Wheels Turning: 15 May – August 1970**

I was initially assigned a flight slot in the First Platoon, flying slicks. I was also appointed to be the company motor officer, responsible for the maintenance of all the motor vehicles in the company, which included dozens of vehicles (several jeeps, a water tanker, fuel tankers, two wreckers with a hook and crane, deuce-and-a-halfs, ¾ ton trucks, and a large front end loader that didn't run because it had had a flat tire as long as anyone could remember).

When Sechrist appointed me motor officer he made it clear that the motor pool operation was in bad shape and the overall condition of the company's vehicles reflected it. He wanted all this fixed, and he wanted me to do it — YESTERDAY! The tone of his voice implied that somehow I was responsible for the mess the motor pool was in, but the only logic I could find for his attitude was maybe he was mad because I had not been there earlier to clean it up.

In reality, the only thing I really knew about motor vehicle maintenance was what I had picked up on my own watching others, or doing minor maintenance on my own cars. I knew the assignment would be challenging, with almost no training to fall back on. But I did have good organizational skills and a very stubborn will, which turned out to be keys for success.

The motor pool turned out to be a bigger problem than I anticipated. To start with, there were almost no replacement parts on hand, even though the company was authorized to maintain a basic level of parts stock. Even the most basic things, like spark plugs, tires and batteries, had to be ordered from the supply unit in Saigon, 100 miles away through hostile territory. It would often take days or weeks for the orders to be filled. What was truly amazing, though, was the paperwork that the parts business required. Every part had to be ordered in triplicate copies that could only be made using a typewriter and carbon paper (relic technology left over from the first half of the 20$^{th}$ century).

The second major problem was the tool inventory, which was mostly empty. There were half a dozen dirty, greasy tool boxes, but other than a pair of pliers or a hammer or screwdriver here and there, little else. When asked, no one seemed to know where all the tools had gone, and by the looks of genuine bewilderment on the faces of the mechanics, I believed them. As they told it, there were no tools on hand when they had arrived for duty and no one had been interested enough to get new tools on order. Plus there was no security or sign-out procedure for larger or more expensive tools.

So the first order of business became fixing the tool shortage, which directly involved the motor sergeant and a specialist 5$^{th}$ class named Pomona, who was a city-raised Chicagoan about my age. Specialist Pomona was the ring master of the motor pool and the informal leader of the crew. His most distinguishing characteristic, besides the ever-present sunglasses he wore, was his attitude: he was a wiseass, but in a nice way. He was a Midwestern wiseass, someone who could make benign wisecracks that had no malice behind them and were just oblique enough not to raise any real hostility. Pomona's manner was quite unlike the downstate New Yorker wiseasses I had come across from time to time, people whose remarks often displayed genuine disrespect and contained threatening tones of ill will. As far as I could tell, Pomona was a nice guy and, like everyone else, I grew to like him.

Pomona and most of the others were apologetic about the motor pool and I could tell that the poor condition of it bothered them. To get started I directed that a couple of sheets of plywood be mounted on the inside wall of a large trailer that could be locked for security, and then directed the outlines of various tools be drawn on the plywood so that when new ones arrived they could be hung in the place where their outlines were, and everyone would know by its absence when one was missing. This assignment was greeted with enthusiasm, but also a slightly detectable air of doubt that this exercise could make any real difference. However, once the new tools arrived and new accountability procedures implemented, we never had a problem with disappearing tools after that.

Another problem was the lack of an office in the motor pool, so a makeshift one was set up in the back of a metal trailer that had one open end, but was otherwise completely enclosed. During the day the scorching sun would heat the trailer like an oven and it was impossible to stay in it to prepare requisitions or other paperwork. Even in the early morning before the sun reached its highest intensity, you couldn't sit in the trailer without sweating gallons, which would eventually run down bare arms and drip onto paperwork, sometimes smudging the words and requiring do-overs.

There were several other mechanics assigned to the motor pool. One of them was a SP4 named Earnest, a man of medium build with brown hair who was rather quiet and slow mannered, but not stupid. He was a good mechanic who did his job with no complaints. He didn't talk much himself, but he liked to hear others tell jokes. When he was frustrated or confused, a scowl would appear on his face that made him look like a little boy who had just been told he was having brussels sprouts for dinner. Earnest was easy to be around and didn't seem to be either impressed or intimidated by rank. We got along well.

Another mechanic was a SP4 named Farkas, a freckle-faced, red-haired fellow from deep in the Appalachians. Farkas was slower in both manner and wit than Earnest, but with none of his redeeming qualities. He was about 5'10" tall and was pudgy around the middle, so going shirtless to keep cool was not becoming. When he laughed he cackled. And he seemed inclined to do as little work as possible, although he would make a good show of what he did do by getting his hands as dirty as possible. Farkas often attempted to emulate Pomona's wisecracking demeanor, but lacking both the intelligence and simple charm of his mentor, or the genuine hostility of a downstate New Yorker, he instead came across as a buffoon.

There were a couple of other mechanics, late arrivals as I recall, whom I don't remember well, named Philbert and Johanson. But the last motor pool character I do remember well is Papa San, whose real name was unknown to all of us.

Papa San was a Vietnamese mechanic, no more than five feet tall, who was hired by the U.S. government to work in the motor pool. He was probably in his 50s and spoke almost no English (at least as far as I knew). He came to the motor pool on foot everyday from his home somewhere outside the main gate of the airfield. Papa San was a quiet man with a seriousness about him that was sobering. I don't recall seeing him smile much, but he was respectful of every one and trusted by all. As far as I could tell, respect and trust were the essence of the man. (One day he brought me a pineapple to eat that was as sweet as candy, a sign of real friendship and respect.) Small of physical stature, his second-hand GI clothing was vastly oversized and he always looked like he had crawled inside a larger man's uniform. While most Vietnamese wore rubber flip flops, Papa San proudly wore combat boots that seemed to be many sizes too big for his feet, and they flopped heavily when he walked.

There were quite a few local Vietnamese who worked at the airfield as mechanics, barbers, masseuses, maids, waitresses and general utility employees. How they were separated from genuine Viet Cong (VC) or VC sympathizers and cleared to work at an American military installation was completely unknown to me. I just assumed their presence was part of the routine that had been set by our predecessors and didn't think much about it, except when the airfield was under a night mortar attack by the VC. Then one couldn't help but wonder if some of the employees were providing tips to the VC on where to aim their mortar tubes. I never had any reason to suspect that our motor pool Papa San was either VC, or a VC sympathizer.

**Carb 6**

As one might expect, the essence and pride of an aviation company were its aircraft and air crews, with pilots being the "most glorious" of all, and air missions being the reason for its existence. Operations like the motor pool and the supply office that were necessary to support air missions were considered secondary in importance, and therefore in prestige. This perspective produced a general pecking order in the company, with air crews at the top and everything else and everyone else somewhere below that.

Near the very bottom of the pecking order were the motor pool and mechanics. We comprised an essential operation within the company that was recognized only when mission critical vehicles weren't where they were supposed to be when they were needed. It was as if the best one could do in the motor pool was stay out of trouble, and there certainly was no glory in that.

The motor pool personnel reflected this perspective, demonstrating their lower class mentality in what they did and said. Continually on the search for ways to raise their self image, they talked about flight crew members with resentment and jealousy, mostly through a sarcasm they had finely honed. My approach to fixing the motor pool operations was not limited to just parts, tools and procedures; it included "fixing" the psyches of the men by providing leadership that I thought would raise their self-esteem and self confidence.

Just off the cuff one day, Al jokingly said my radio call sign should be "Carb 6" (carb being short for carburetor, a part of the engine that metered the flow of fuel in "ancient" motor vehicles). This idea got around, and pretty soon just about everyone was calling me Carb 6, including the guys in the motor pool. A few days later, "Carb 6" was stenciled in white letters on the front of the olive drab motor pool jeep just below the front window, in exactly the same place that "Boomerang 6" was stenciled on the company commander's jeep. When Pomona and the boys showed me this handiwork, of which I approved, there was considerable enthusiasm in the group. In their eyes this small act raised my stature to that of the company commander, which in turn raised their stature to at least that of the fly boys who got all the attention. Even Papa San seemed to sport a new self-esteem.

To me, this minor horse play was good stuff. It raised morale at a time and in a place where high morale was not easy to come by, and it didn't do so by resorting to drugs, alcohol, sex or violence. Unfortunately, the company commander didn't see it that way. A few days after Carb 6 came into being, I was summoned to Major Sechrist's office where he proceeded to reprimand me for driving around in a jeep marked "Carb 6." He especially disliked my driving the jeep between the motor pool and the officer's quarters, a trip I made several times a day. According to Army-based reasoning, junior officers such as I were not "authorized" to use Army vehicles for personal transportation. Oh yes, and he also wanted "Carb 6" removed from the vehicle.

This directive was not unexpected because I figured Sechrist to be a "lifer" who lacked the ability to grant esteem if it might be at the expense of his lofty stature. I could accept that perspective and his directive, because I knew I might have one toe over the line of my authority as a lieutenant. So I resigned myself to what seemed like the inevitable outcome of the Carb 6 affair.

But I wasn't at all prepared for what came next. As I stood before his desk, Sechrist reached back and grasped the edge of a manila folder that was taped to the wall behind him. He flipped the folder open revealing a colorful cartoon of a clownish-looking GI who was standing on his own penis with a facial expression of combined pain and embarrassment. The caricature implied that I was in the same position as the cartoon GI was.

Sechrist grinned. I was chagrined! I didn't think this little joke was in good taste, and it certainly wasn't becoming of a U.S. Army major. Nor did it respect the hard work and dedication I was putting into the motor pool assignment. Sechrist had revealed himself to be crude and arrogant, a hack with a gold leaf on his collar who had proved he could not be trusted to show respect or provide support.

I returned to the motor pool disheartened. We had something good going for us there, and I didn't know how to deal with this turn of events. After a brief explanation of what had happened I told the men to remove the "Carb 6" stencil from the jeep. I know it hurt their pride to paint over the letters, yet even though the paint covered the words it was clear these men could not return to the old order. They knew now that I was willing to go to bat for them and take heat in their behalf, that I considered them worthy of such effort. I had become one of them, and one with them - the essence of leadership.

**Buzzards**

From that point on, the motor pool crew worked effectively as a team. New tools arrived and were put into service, a reasonable supply of spare parts took shape and the overall condition and reliability of the company's equipment improved dramatically. But in the midst of all this progress I was awakened about 3:00 AM one dark morning and told that someone had been caught trying to break into the motor parts storage container. I grabbed my pants and shirt and ran to the motor pool just as the MPs were putting the thieves into a jeep to cart them off to the clink.

The suspects caught in the act were GIs assigned to our "sister" company, the 162nd AHC (the "Vultures") whose motor pool was next door to ours. I was angry and disappointed to think that our fellow GIs would stoop to theft to get what we had come by through hard work, good organization and determination. I was naïve on this subject and had a lot to learn about what people will do when they are under stress or in difficult situations.

In August, following three and a half months of responsibility for the motor pool, my duties were turned over to 1LT Duane Henderson. I was mentally relieved and glad to leave the motor pool behind me, but I was also glad to have known the men and to have learned more about leadership from Pomona, Earnest, Farkas, Philbert, Johanson, and Papa San, too. We had accomplished much together and the only regret I had in leaving was that in spite of all the hard work, one task remained unfinished: the flat tire on the front end loader was ---------- still flat!

(Letters I wrote to my wife at the time reveal that Sechrist was under serious pressure from his higher-ups about the miserable condition of the vehicles in the company, which may explain why he was such a jerk in dealing with me. Only a few vehicles were operational out of the 35 that were in the company when I took over the motor pool, but almost all were operating when Henderson took over three months later. Moreover, we had passed a critical inspection that saved Sechrist's butt, and probably the evaluation of his command performance too. I did what many of my predecessors failed to do, but got nothing from Sechrist for it, not even an "attaboy".)

## Boomerang One-Zero

The motor pool occupied roughly one-half of my waking hours during the first three and a half months of my tour of duty. While it was not exactly dull, it lacked the appeal, the prestige and the emotional rush that came from flying slicks during the other half of my waking time.

Upon arrival at the 191st on 8 May, I was assigned to the 1st Flight Platoon as the 1st section commander, and was given the radio call sign Boomerang One-Zero. Now, having a title like section commander may sound impressive, but in fact I had no idea what my duties were (and no else seemed to know). I also had no idea who was assigned to my section! It was essentially a formality as far as I could see, because the real authority in the platoon, and in the entire company for that matter, was held by the experienced pilots who knew how to fly a Huey under fire. In combat conditions everyone deferred to the experienced pilots because they usually knew how to stay out of trouble, and that was one of the key factors in going home alive. On 8 May, everyone in the company had more flight time than I had, so I was the low man on the totem pole and I deferred to everyone, all the time, even the enlisted crew members. Rank just didn't mean much at that point.

## Boomerang One-Six

The 1st platoon commander was a captain named Ron McConkey. He was about 35 and had risen through the enlisted men's ranks to eventually attend Officer Candidate School (OCS) and flight school. Nearly six feet tall with a wide forehead, dark hair and an overbite that produced a slight lisp, he was even tempered, good natured and calm. He smoked heavily, but drank alcohol moderately.

McConkey was on his second tour of Vietnam, having spent his first tour as a military adviser. He was friendly and unassuming, patient and unflappable, taking almost everything in stride. He seemed to be a textbook example of a man who had risen from the enlisted ranks to become an "officer and gentleman." He'd also been around the Army long enough to know how to stay out of trouble.

Experienced men like him understood the subtleties of the Army's system. He knew there was a distinct line between what the system regarded as acceptable and unacceptable, and just how close he could get to that line and still be safe. He also realized that ambition could be punished just as severely as bad behavior, and maintained a low profile to avoid negative repercussions from anyone who could affect his career.

McConkey's leadership style was laid back; he didn't push his men into anything. Instead, he let them seek and find their own levels of comfortable performance, opting into a friendly "partnership" with the people in his platoon. Everyone seemed to like and get along well with him.

One other thing about Ron McConkey - he had an intense dislike for the Vietnamese people, and always referred to them as "gooks." It seemed that in his mind there was no difference between South Vietnamese civilians or soldiers, VC, NVA (North Vietnamese Army), or North Vietnamese civilians. When the subject was the Vietnamese people, he was an equal opportunity bigot; he hated all of them.

Occasionally McConkey liked to go downtown to the city, a mile or so away from the airfield. But the only safe way to walk there was by way of the always-busy, two-lane road that ran by the front gate, a road that was safer in daylight than at night.

During the day the road was always packed with people, animals, ox carts, trucks, cars - and "cowboys." Vietnamese cowboys were young men in their teens and twenties who drove motor bikes along the road in packs of 5 or 6 looking for "targets of opportunity" to terrorize or steal from. They were essentially outlaws who hated Americans with a passion equal to McConkey's.

One of the cowboys' favorite targets was any GI walking along the road who was wearing a wrist watch with an expandable band. As a cowboy approached a GI from the rear, he would time the approach so that he passed by just as the GI's arm nearest to him would swing forward. As he passed his victim the cowboy would quickly slip his fingers between the GI's arm and his watch, and the forward motion of the motor bike would cause the watch to slip over the GI's hand as the watch band expanded. Then the pack would zip off into the crowds with their new trophy, making it impossible to stop them.

There were many reports of such incidents, and McConkey himself had been victimized that way. But one day, fed up with the situation, he got his revenge. He went to the Post Exchange (PX) and bought a new watch with a solid band that looked like it was expandable, and walked downtown with it on his arm. As expected, a cowboy zipped by and slipped his fingers between the new watch and McConkey's wrist, but this time the band didn't expand. The cowboy was jerked off his bike and

McConkey took great delight in cleaning up the road with him. (This incident didn't help international relations at all, yet one couldn't help feeling that the cowboy had it coming.)

## Cleared to Fly

Wednesday, 13 May was the first day I took off in a Huey from Can Tho Army Airfield. It was exciting, fulfilling and unsettling all at once as Dave Armstrong, who was the company instructor pilot (IP), put me through a check ride and orientation procedure that would qualify me to fly combat missions with the 1st Platoon. With much uncertainty in mind about Dave, whom I had met just a few days before, and the hostile terrain we were flying over, thoughts of failing to find Don Scott Field as a student pilot at Ohio State came flooding back. In this case, primarily because of Armstrong's calm demeanor, I managed to fight off the fear of failure. (It would not be the last time this would happen.)

The most challenging part of the hour-and-a-half ride was at the end, trying to park the Huey in its narrow concrete revetment at Can Tho Field. This was something that could be challenging even for the most experienced pilots because of cross winds, or other Hueys hovering in the area that could create unpredictable wind currents. But I did it, and after the paperwork was finished the check ride was pronounced a success. After many months of preparation I was finally cleared to fly as a combat pilot, ready to take my place in the Boomerang flight.

From a letter written on 13 May 1970:

> *This afternoon I took my check ride. Went up the Mekong River to a little paved strip*

*and did the usuals-- autorotations, normal approaches and takeoffs, emergency procedures, etc. It went real well-- guess I haven't forgotten how to fly after all. After that we cruised around and looked at Can Tho Airfield and Can Tho City from the air. This certainly is beautiful country over here-- from the air everything is lush and green, except for the plowed fields and rice paddies. The paddies are all dry now, but the monsoons are starting and in a month or so they will all be under several feet of water. The Mekong River is huge and has lots of big islands scattered through it. And the cloud formations are something else. Never seen anything like them before, anywhere. You get big black storm clouds and light cumulus all mixed together. I can't really explain it. Will have to take some pictures and send them along.*

Armstrong, from Oregon, was quiet, serious, intelligent and highly skilled, one of the best pilots I ever flew with. His sandy colored hair augmented his good looks, but his large drooping mustache made him look like Pancho Villa (a nineteenth century Mexican revolutionary general). When I arrived at the 191st, Dave had about eight months' combat flying experience which had given him an intimate understanding of the Huey's capabilities and limitations. I flew missions with Dave a number of times during my first four months as a Boomerang, and although I was petrified on more than one occasion I had no reason to doubt either his skill or his judgment.

**Chau Doc (*chow duck*)**

On Sunday, 17 May, I was up at 4:00 AM to prepare for a 5:15 takeoff on my first combat flight. After getting breakfast in the mess hall, I joined other Boomerang pilots at the 191st operations office to receive a briefing on the combat mission for the day. The operations officer told us we were to rendezvous with ARVN infantry units, and several other flying outfits at 6:00 AM and lift them from the PZ near Chau Doc into Cambodia (this was the early stage of President Nixon's famous and controversial "Cambodian incursion"). Several sorties (round trips from a PZ to an LZ) would be required to move all the troops. We were given the map coordinates of the PZ and the names and call signs of our infantry counterparts who would be on the ground waiting to load their troops when we arrived.

It was completely dark as we walked to the flight line, but a hint of dawn was starting to color the eastern horizon. Even at that hour of the day the air was balmy and tropical, the overnight temperature bottoming out at about 75°F. But we knew it would top out near 95 by early afternoon, with 80-90% humidity.

In the limited glow of a flashlight I did a preflight inspection on the Huey while the pilot in charge (PIC) talked with the crew chief and checked the flight log for any mechanical problems. We could hear the four other slick crews laboring in the dark getting ready for takeoff, while the two gunship crews cranked up their Hueys and headed for the rearming point near the flight line. We cranked our Huey with the rest of the slicks and waited briefly for the guns to finish arming with rockets and machine gun ammunition needed for the M-60s that pointed out the two cargo doors of each gunship.

At 5:15 the entire flight lifted to a hover and moved into position in a long single file on runway 26. The air was completely calm, but the rotor downwash from the Hueys was strong and it swirled the oily, burnt smell of JP-4 fuel up our noses and around our flight helmets, making our eyes water. The Can Tho control tower operator gave the flight clearance to take off and we moved forward, gradually gaining speed until we felt translational lift, and the Huey leaped up and forward as a result of fresh air flowing through the rotor blades. It was a phenomenon we had all grown used to from the early days of flight school, and it was now a benchmark that foretold a successful takeoff.

We climbed at a steady rate to our standard cruising altitude of 1,500 feet above the ground (an altitude just out of small arms range) and formed into our standard V flight with the gunships cruising beside us, one on either side of our formation. The flight turned northwest and began following the massive Bassac (*bah-sock*) River that would lead directly to Chau Doc. At 1,500 feet the air was clear and refreshing, and by now the dawn was underway. The terrain, flat as a pool table beneath us, was covered with clearly defined rice paddies of shallow glassy water just deep enough to reflect the dawn's rays. We flew over small, sleepy villages on the edges of canals that joined the Bassac where gray colored sampans were tied up to the shore, and over an occasional narrow dirt road lined with palm trees. Small military outposts could be seen here and there if one looked closely and long enough. At our altitude it was serene and beautiful, a land of mirror glass and deep green. But there was no escaping the ceaseless roar of the Huey's jet engine in the background or the occasional radio message between the pilots in the flight that overshadowed the panorama.

After forty-five minutes of flying the sun peaking above the horizon made us squint as we broke formation and lined up for our final landing approach to the road that was our PZ. ARVN troops in olive drab uniforms carrying M-16s were visible next to their trucks at the PZ. As we descended to land, our two gunships flew beside us to provide fire support in case we started receiving hostile gunfire, but in this case all was quiet. At my PIC's insistence I had been flying since taking off just to gain flying experience. But once we had landed and were at a hover he took the Huey's controls and set it down at the rear of the flight, just off the runway.

There were numerous other helicopters from American units and the VNAF (Vietnamese Air Force) parked nearby and ARVN troops came to board soon after touchdown. Significantly smaller than Americans, about a full head shorter, a Huey would hold 11 or 12 ARVN soldiers whereas only 6 or 7 Americans would fit into the same space. We waited a few minutes as the ARVN infantry commander talked with the commander of our flight. Everyone seemed focused and loose. I was apprehensive, puffing nervously on a Marlboro.

About ten ARVN soldiers with M-16s, M-60s, backpacks, radios and other gear piled into the cargo area of our Huey. Our crew chief and door gunner were squeezed uncomfortably against their M-60s, but they didn't complain. They had been through this many times before and knew what to expect. Soon we were in the air with numerous other Hueys, all headed into Cambodia. We dropped the ARVN onto open ground and headed back to Chau Doc where we waited for additional instructions, but I cannot recall if we made any more insertions that day. What I do recall is how much confusion there was. No one seemed to know what was supposed to happen, or how or when, and that's how the whole day went until late afternoon when we returned to Can Tho Airfield (CTA).

Several pilots commented that the confusion we had experienced that day was not typical of a 191st mission, and blamed it on the large number of separate units that were involved. I expressed surprise at the ineffectiveness of the effort, but they assured me that most combat assaults were far better organized and conducted when we were not involved with other units. After today, I was looking forward to better missions.

From a letter written 19 May 1970:

> ***Sunday morning while we were on standby status during the combat assault, we remained in the staging area which was simply a large rice paddy. All the local kids from around the area came up to the helicopter and started looking around. They are really curious, as are all children. I gave them all my gum and it was very interesting trying to communicate with them. About mid-morning a young Vietnamese about 15 or 16 came up to the group and started watching. I noticed he had a piece of paper in his pocket and asked to see it. It had 5 or 6 phrases written out in both English and Vietnamese on it. We started talking back and forth, me teaching him the English and he teaching me the Vietnamese. We became best of friends in about an hour. It certainly is surprising how much warmth can be transmitted even when you cannot talk with someone. When I got out of the Huey, he***

*followed me around and one time I noticed he had his arm around my shoulder.*

*A Vietnamese soldier also came up to me and we talked a little. His name was Quang and he must have been 20 or so. He was very interested in a plastic cover I had on my dog tag chain, so I gave it to him - he was extremely grateful.*

This experience, although early in my tour, helped to confirm what seemed obvious, that success in this war was not possible. In the letter referenced above I also wrote:

*Dave Sturgis, the warrant officer that went thru school with Tom Brown, flies the ARVN General around that controls the IV Corp area. He says that the ARVNs can't possibly lose after we leave. But, I believe he's wrong. As soon as we leave, Charlie (i.e. the Viet Cong) will have his own way.*

And in another letter dated 22 May 1970:

*I've come to the conclusion that this whole thing over here is the most ridiculous damned thing that the U.S. ever got itself into.*

--- Ω ---

On Sunday, 24 May, we left CTA at 7:00 AM and headed almost due north to the small town of Moc Hoa (*muck wah*), a stone's throw from the Cambodian border near the "Parrot's Beak" (the

Parrot's Beak is an area of land that belongs to Cambodia that juts southeastward into Vietnam). Like my first mission a week earlier, this mission was to insert ARVN troops into Cambodia. This time, however, the 191$^{st}$ worked alone and I got to see just how efficient and effective the Boomerang pilots and crews were.

The first insertion was smooth and uneventful, and we returned to Moc Hoa for a second lift of troops. However, the troops in the second lift were dropped on the opposite side of the LZ from where the first lift had been dropped, and because of some apparent miscommunication the first and second lift troops began shooting at each other! It was a mess, and even though it was a close call, we got out of the LZ quickly and safely.

In the afternoon we lifted the ARVNs back to Moc Hoa and then returned to CTA. The only action that had been encountered the whole day was when the ARVNs were shooting at each other in the LZ. It looked like more evidence of the incompetence that supported the belief they could not win the war without our continuing help.

**Magnet Ass**

A few days later we were back in the air at 1,500 feet in our standard V formation, our Huey again in right trail position en route to an LZ. The friendly radio chatter between the pilots in the flight going to the PZ was gone, replaced by a cold silence and an occasional message from the lead pilot in the flight. I was navigating (following the map) and Dave Armstrong was flying. His insistence on overlapping our rotor blades with the Huey in front of us was unnerving, and I asked him why we were following so closely. He said it was a matter of control and comfort; that the air flowing through the rotors of several Hueys that close together made for smoother flight, but it was certainly not comfortable for me. It seemed like an unnecessary risk, but I didn't say anything; I just thought to myself "I'm not going to like this formation flying business if this is always what it's like."

Ten minutes later, the lead pilot gave the flight instructions about how to make our approach to and exit from the LZ, and we started our descent. The Bounty Hunter pilots on each side of the flight began firing their rockets into the LZ and their M-60s into tree lines leading to it. The ARVNs in the back were preparing to jump out as our guys started firing their M-60s into anything that looked like a target. The air became choppy and our Huey started bouncing up and down in the turbulence, our seat belts straining from the extra gravity load. As my legs went rigid with fear I could see Armstrong had his hands full. I was completely transfixed by what was happening.

As we got closer to the ground the air got warmer. Sweat poured into my eyes, the stinging from it adding to the anxiety of the moment. Rockets were transforming the LZ in front of us into clouds of earth, grass and smoke, and as we got closer the concussion from each one shook the cockpit. I could faintly hear the spitting of our M-60s above the whine of the engine. Exhaust fumes from the Hueys just a few feet in front of us were streaming through the cockpit and into our lungs, and I struggled to find oxygen in the ceaseless stream of spent aircraft fuel. Armstrong was like a machine, a cold mask across his face showing nothing but intense focus on the task. I was nearly frozen in my seat, useless.

Our wedge of five Hueys had descended almost to tree top level when the VC opened up on us with small arms fire. We didn't know it until the Bounty Hunter pilots saw the flashes of VC rifles from a tree line and announced it over the radio. It was coming from our side of the flight, and on my side of our Huey. I heard several rifle rounds thump into the fuselage nearby, but I couldn't tell where. "Oh, Jesus," I thought, "this is worse than I imagined!"

I looked around and saw none of us were hurt, but Armstrong was grim, his concentration unwavering as the flight came to a low hover. Our M-60s had quit firing and he told the crew chief to have the ARVNs "un-ass the aircraft." With the crew chief pushing them, the ARVNs jumped into waist high grass and struggled across the LZ with their M-16s answering the VC rifle fire. Waiting there at a hover we were as vulnerable as we could be, and all I wanted was space between us and the LZ as quickly as possible.

The Boomerang lead pilot asked if we were empty; Armstrong replied, "Affirmative," and we started moving forward with the rest of the flight. When we hit translational lift I looked out the cockpit window as our M-60s resumed firing on the tree line where the VC were. The Bounty Hunters were pouring rocket rounds and machine gun fire into the tree line too. And as we climbed and turned back toward the PZ, I lost sight of the action.

Several more insertions that day were quiet and in the afternoon we reversed the process and extracted the ARVNs from the LZs, dropping them off in the PZ where they had originally been picked up. We headed back to CTA and when our flight had landed and the Hueys were parked in their revetments, the pilots began to roam from one aircraft to another to examine the damage each one had taken from the VC rifle fire that morning. Ron McConkey wandered over to our Huey to look at the holes from the bullets we had taken, the ones I heard hitting us during our first insertion that had hit the structural pillar just behind my head. Poking his finger into one of the holes in the sheet metal he turned and looked at me. A broad smile crossed his face and he said, "Lt. Barth, you attract enemy gun fire; from now on I'm gonna call you 'Magnet Ass'!"

**ROE**

Eliminating the communist enemy in order to secure a democratic form of government in South Vietnam was the stated overall purpose of the U.S. in the Vietnam War (or so I thought). It was a difficult undertaking, complicated by our inability to locate, fix and destroy the enemy who came in two versions: North Vietnamese Army (NVA) soldiers who were uniformed and organized into identifiable military units; and Viet Cong (VC) who dressed like civilians, lived and worked among the general population, and were organized in small cells or units and did not present themselves openly. They were harder to locate and destroy.

In the Mekong Delta there were relatively few NVA, but many VC who blended seamlessly into the local population. Their tactic was to remain hidden until a moment of their choice when they would strike against American or ARVN forces quickly, do as much damage as possible, and then withdraw just as abruptly to again blend in with the population or the countryside.

In order to control tactical operations and prevent the innocent civilian population from being destroyed along with the VC, a system of rules for American troops was invented called the Rules of Engagement (ROE). No one really seemed clear about what the rules were, but we all knew they existed as ironclad restrictions about identifying and engaging the enemy. Generally speaking, any Vietnamese dressed in civilian clothing and carrying a rifle was considered to be the enemy.

Enforcing the rules during our combat missions was the responsibility of the ranking officer in the command and control aircraft (referred to as C&C) that flew with us at 1,500 feet above the ground during all combat missions. When we were flying combat assaults, the C&C was almost always comprised of our company commander and a ground unit commander, or ground counterpart. This two-man team directed the flight by radio and determined who we could or could not engage.

Because C&C was at 1,500 feet and not ground level where the slicks operated, we often saw things that the C&C team could not see, such as black-pajamed VC who would drop their AK-47 rifles under water in a rice paddy and act as if they were ordinary rice farmers to avoid detection. We slick pilots knew they were probably the enemy who had just fired at us when we were landing in a nearby LZ, but we were forbidden to fire upon them because they appeared to be unarmed to C&C.

Unfortunately, the VC knew about our rules of engagement and used them to their advantage, and our knowledge of their practices produced great frustration for us Americans, even as it protected the innocents (some of the time anyway).

## Initiation

One would have expected that flying into combat for the first time would have constituted an initiation, but that was apparently not sufficient to satisfy the more juvenile members of the Boomerangs. So after dark one night a couple of weeks after I arrived at the 191st, some of the more experienced pilots organized a handful of us new pilots ("newbies") for an "initiation." We were blindfolded and directed to drink some unknown alcoholic concoction, then paraded outside to consume a piece of a leaf from a "magic plant" as it was called. The 3 or 4 of us being initiated dutifully ate the leaf, which seemed to please the initiators a great deal.

I had expected the leaf to be from a marijuana plant, or some other type of drug containing plant. But when the blind folds were removed the initiators disclosed that the plant was an ordinary banana tree that had been urinated on numerous times prior to the initiation. I suppose to most people this would have been disgusting or revolting, but to me, having spent time growing up on a dairy farm where handling cow manure was routine, it was just juvenile and stupid.

## Can Tho City

Can Tho City is located in the center of the Mekong River Delta at the junction of the Can Tho and Bassac Rivers. The Bassac is the southernmost branch of the Mekong River, which originates in the Tibetan Plateau in China and flows over 3,000 miles south and east through Burma (Myanmar), Laos, Thailand, Cambodia and Vietnam, eventually emptying into the South China Sea. In 1970, the city was large with a population of several hundred thousand, but relatively backward compared to American standards. The city was in Phong Dinh (*fong din*) Province which bordered six other provinces: Ba Xuyen (*bah zwee-en*) to the southeast where Soc Trang was, Chuong Tien (*chung tin*) to the southwest, Kien Giang (*ken zhee-ang*) to the west, An Giang (*ahn zhee-ang*) to the northwest, Sa Dec (*sah deck*) to the north, and Vinh Long (*vin long*) to the east.

Can Tho was heavily influenced by western cultures, particularly the French who dominated Vietnam (formerly called Indochina) as a colonial power prior to the 1950s. The French lost their control through military defeat by Viet Minh rebels (later the Viet Cong) in a decisive battle at Dien Bien Phu (*dee-en bee-en foo*) in 1954. French companies sourced much of their raw material supply from plantations throughout the Delta. Several centuries of French influence could be seen by the existence of Catholic schools, French architecture and people of mixed French and Vietnamese blood lines. But, the relatively advanced culture of Can Tho City was very different from the rural areas where the VC still had strong influence and control.

My expectations of Vietnam had been built on the field training we experienced in flight school, which took place in a crude setting in the pine woods of Georgia. There we had bunked in a wooden cabin perched on the edge of a small grassy clearing just big enough for several Hueys to land and park in. We ate cold C rations. There was no running water or electricity. No radio, no phones, no entertainment. I thought this was what war duty would be like. Can Tho Airfield, however, was nothing like that.

**Can Tho Airfield (CTA)**

CTA was about a decade old and had been continually upgraded over that decade to produce a secure, relatively comfortable base of operations. It was like a chunk of the USA had been plunked down in the middle of Phong Dinh Province, with nearly all the comforts of home.

The airfield had one runway about 3,000 feet long that was laid out almost directly east and west, constructed of metal landing strip material called PSP (pierced steel planking). Each strip of PSP was about 1.5' wide by 12' long and hooked to other pieces of PSP to make a runway surface that simply lay on top of the ground. The runway was flanked with water-filled, rectangular ditches several feet deep, created when the airfield was built. In order to get enough soil for a dry runway, the earth had been excavated and mounded into a long strip which left open ditches next to the runway. Because the Mekong Delta was barely above sea level, almost all depressions deeper than a foot or two automatically filled with water.

Located about a mile and a half northwest of downtown Can Tho, the airfield had a tower that was operational day and night to control air traffic in and around the airfield. Other than that, there were no navigational aviation aids and most flying was done under VFR conditions (Visual Flight Rules). Occasionally pilots would have to use instruments for takeoff and/or landing when the weather prevented VFR (known as Tactical IFR), but that was relatively rare. I recall intentionally entering Tactical IFR only once during my tour.

The runway and taxi areas were surrounded on three sides by aircraft parking areas, maintenance hangars, billets for enlisted men and officers, unit headquarters buildings, operations offices, mess halls, motor pools, sand bagged bunkers and many other structures. Most of the service facilities were on the north side of the runway, including a post exchange (PX), a massage parlor (where more than massages were offered, so I heard), a barber shop, a laundry, a radio station for making phone calls home, an enlisted men's (EM) club, an officers (O) club, a medical and dental dispensary, and the offices of the 164$^{th}$ Aviation Group which were at the northeast end of the runway. There was even a POW compound on the northernmost edge of the airfield.

The roads between the buildings were unpaved, but sprayed with oil to reduce the dust. Most of the buildings were standard Army style, wooden one or two-story structures like the buildings I had first seen at the 90$^{th}$ Replacement Battalion at Long Binh. While the roads were unpaved, there were concrete sidewalks in many places.

The PX offered beer, cigarettes, a limited selection of canned food, clothing, some electronics products (like Seiko watches, which were deemed necessary to buy before leaving), electric fans, and books and magazines.

There was a reasonable amount of entertainment available with USO tour shows offered now and then at the EM and O Clubs, and there were movies available when there were no live shows. For the most part USO shows were almost always rock and roll bands from the U.S., the Philippines, or Australia that covered the popular music of the times. But after seeing a few of them I realized they all had the same format: three or four guys on instruments and one or two go-go girl singers dressed in skimpy bikinis and boots. Armed Forces Vietnam radio (AFVN) was available everywhere and we often listened to it on the Hueys' radios when flying to keep up with new pop music in the States.

There were even Vietnamese and AFVN TV stations available for anyone who had a TV and a decent antenna.

The weather at Can Tho was tropical with two main seasons: monsoon (or rainy) and dry. Monsoon season started when I arrived in May and ran through November, with rain nearly every afternoon and daily high temperatures above 90°F. The rain and high temperatures brought high humidity with them (the average humidity in Can Tho was 83% year round) resulting in constant perspiration and soaked clothing. The only relief was in an air conditioned building, like the officers club, or at 1,500 feet above the ground where the temperature was around 70°. Dry season from November through May was just as tropical but with less rain. Temperatures were only slightly cooler and the humidity was a little lower, but in the daily routine the differences between the two seasons were almost irrelevant.

Security was always of concern to us foreigners in a foreign land. The perimeter of the airfield was secured by several rows of concertina wire, Claymore mines and bunkers for guards during the night hours. This mix proved to be effective as there was never a breach of the perimeter by the VC while I was at Can Tho. Yet the VC could get access to the airfield with indirect fire, using mortars to lob rounds from outside the perimeter onto the airfield. During my tour of duty we were mortared about once a month and whenever the first round hit we would head for the nearest bunker a few steps from our bunks, and only a few more steps from the O Club.

The bunkers, while effective protection, were not places where one wanted to spend much time. They were essentially ditches in the ground, about 20 feet long and 12 feet wide, surrounded and covered by sand bags. They were not high enough to stand up in, and the dirt floors were usually wet or muddy, especially during monsoon season. They were damp, musty and claustrophobia-inducing. Not used very often or regularly maintained, one had to be on the lookout for rats, snakes and spiders going into a bunker during an emergency.

The VC who mortared the airfield had to act quickly because there would be a Huey in the air looking for their position within a couple of minutes of the first round going off. Usually they only had time to fire a few rounds, so they would select targets that would cause the most damage. The primary targets were the aircraft at the field, including Hueys and some fixed wing planes on the north side of the runway, and Chinooks (twin rotor helicopters) and fixed wing planes on the south side. One night they launched a few rounds into the enlisted men's quarters near the flight line and wounded a couple of GIs, and another night the officers club sustained some damage.

Many Vietnamese girls worked at the airfield, and they had a few sayings they used repeatedly in their banter with Americans. One was "You numbah tay-unh" (you number 10), meaning you were very bad, or "You numbah one," meaning you were top notch in their estimation. Another saying they used often was "Fo' sue-ah," meaning "for sure". Being stationed at Can Tho Airfield was "fo' sue-ah" luxurious compared to what I had expected, or to what my infantry counterparts experienced in the field. Even in the midst of extreme stress, I was always grateful to be a Huey pilot and not have to experience combat in the swamps and rice paddies of the vast Mekong Delta.

# Chapter 2 - JUNE 1970

Four weeks down; only 48 more to go.

| *The Long and Winding Road* - The Beatles |
|---|

Early in the morning on 1 June we received about twenty VC mortar rounds, four of which tore two-foot holes in the 191st maintenance hangar roof and peppered the five Hueys inside with shrapnel, but no one was hurt. Later that morning I drove to Binh Thuy AFB to see if any replacement parts had come in for the motor pool, and on the way back I saw Tom Yost walking toward the Binh Thuy helipad. He had been to the finance office on an errand as the 121's pay officer and was looking for a ride back to Soc Trang. I gave him a lift to CTA where we had lunch at the mess hall and talked for a few minutes.

Following lunch and finding there were no flights to Soc Trang until 3:30 PM, we decided to return to Binh Thuy where Tom attempted to visit LT Macahilig at 3rd Surgical Hospital, the attractive Army nurse who we met on the flight from Saigon a few weeks earlier. I dropped him off at the hospital and picked him up an hour later. He was pleased that he had found her off duty, but frustrated that he had been able to spend only an hour with her. He said the visit had made his day, but had left him wanting to see more of LT Macahilig.

## Shakedown

On Friday, 5 June, I was up at 3:30 AM to participate in an unannounced inspection of the company's enlisted men's barracks. Such "shakedowns," which all company officers and non-commissioned officers (NCOs) were required to participate in, were scheduled occasionally to recover unauthorized materials that the enlisted men had hidden in their quarters, and it was not popular with them. This was my first shakedown and I expected we would find very little to justify the probable hard feelings this invasion of privacy would trigger.

But I was wrong. Our search that morning turned up a surprisingly large amount of unauthorized stuff including switchblade knives, rifle bayonets, hand tools, one M-16 rifle and a water pipe for marijuana. The pipe was found in the same room where a large cache of marijuana had been found a few days earlier, and the soldier who possessed it was taken off flight status and relegated to filling sand bags, scrubbing the maintenance hangar floor, and scrubbing rust off corroded ammo one cartridge at a time. He reminded me of cadet "Screwup" at ROTC summer camp who could do nothing right, but this screwup had crossed a line that put him on track for a court martial.

The next night I had Ready Reactionary Force (RRF) duty at the company operations shack near the flight line. The RRF was a rotating nightly responsibility to provide a standby security force to repel any enemy attacks on CTA that might occur. Usually it involved nothing more than trying to stay awake all night, and now and then there were mortar attacks that forced a call out of the RRF, but not the night of 6 June, which turned out to be quiet.

A few nights earlier it had been entirely different at a tiny place called Tinh Bien (*tin bin*), 70 miles to the northwest of Can Tho, where a radar station took more than 100 rounds of VC mortar and recoilless rifle fire. The attack destroyed the station, but the GIs held on until they were evacuated the next morning. While I was pulling RRF duty the evening of the 6th, one of the evacuees from the Tinh Bien station was sleeping in my bed. I talked to him before going on duty and he said he was shaking so badly during the attack that he could not hold his M-16. It was a bad situation, and I was glad it happened there and not at CTA.

--- Ω ---

By early June the company's combat assault (CA) missions had dropped off considerably because the 191st was short of Hueys. A half dozen of the Boomerang Platoon's choppers had reached 2,200 hours of flight time and had to be "retrograded" for major inspection and rebuilding, work that was beyond the capability and scope of the company's maintenance team. Another four had been damaged in the mortar attack on the maintenance hangar a few days before and were unable to fly, leaving the platoon with just two flyable Hueys.

One of them had been "red-Xed" (a condition that meant the chopper was not flyable), leaving only one Huey operational in the platoon. So the only missions possible were DCS (Direct Combat Support) missions, or "ash & trash" as the crews preferred to call them, delivering mail, food and other stuff to outposts across the delta. At that point a lot of the 1st Platoon pilots had numerous days off, and no one but the over-worked maintenance crews was complaining.

About this time the company received a new H model Huey, the first one in the company, which before that time had had only C and D models. The new H models were high powered Hueys that flew as well as a new car drives. The prospect of getting more was high, raising the expectation of more CAs in the near future.

**Nighthawk**

Out on the west coast of the Mekong River Delta sits a city of several hundred thousand called Rach Gia (*rock jaw*). The capitol of Kien Giang (*ken yang*) Province, Rach Gia is about 50 miles due west of Can Tho on the shore of the Gulf of Thailand. To the southeast and not very far from Rach Gia is a place called Rach Soi (*rock soy*) which had an airstrip that we used for Huey refueling and rearming.

The 191st had a standing mission to provide air support and protection for the ARVN and U.S. compounds and outposts in and around Rach Gia, and areas to the south all the way to the tip of the Ca Mau peninsula, the southernmost point of Vietnam. Late each afternoon two Boomerang slicks and one Bounty Hunter gunship would fly to the Rach Soi airstrip to standby throughout the night hours in case protection from VC attacks was needed. This was quite different from a combat assault mission, in particular because there were no formation flying or troops involved.

One of the slicks fulfilled the role of C&C and usually flew at 1,500 feet. The C&C pilot in charge (PIC) communicated by radio with the local operations center in Rach Gia to get specific missions, if there were any, and the map coordinates that defined free-fire areas where there was no limit on who or what could be chosen as a legitimate target. Everything in a free-fire area was assumed to belong to the VC. The C&C PIC would then relay this information to the other two ships in the flight and provide overall control of the mission.

The second slick, known as the lightship, had a powerful xenon light several feet in diameter bolted to the floor of the cargo bay and controlled by the crew chief or door gunner. The lightship's purpose was to fly at low level, usually 10-20 feet above the ground, to locate and mark the enemy so the gunship could destroy the enemy position. The job was, quite literally, to be such a lucrative target that the VC could not resist firing at it and thereby disclose their position. In complete darkness the light was incredibly powerful and was capable of disclosing even small objects in deep grass. There was one rather intimidating deterrent onboard that was not the light, it was a .50 caliber machine gun powerful enough to dig into a bunker, pierce armor plating, or rip a man to shreds.

The third Huey was a gunship loaded with two pods of 2.75 inch rockets and two M-60 machine guns. The gunship would circle the lightship as it probed for enemy VC, and when they were found destroy them by any means possible. It was a most deadly mission and a successful one as well, which was shared with no other unit. It belonged exclusively to the 191st.

I was introduced to the Rach Gia night mission very early in my tour, and was more often assigned to this mission while I was in the Boomerang Platoon than to combat assaults. As copilot, I often flew with Chief Warrant Officer Second Class (CW2) Jack Orlo, CW2 Gordon Clayton, and CW2 Rick Mahan. I preferred flying with Mahan and Clayton because they were good natured and showed good judgment, and usually flew C&C. Orlo, with a higher risk tolerance, preferred the lightship.

The standing rules for all night missions were first to be on call at Rach Soi during the hours of darkness. And second, if immediate fire support was not needed, to search for targets of opportunity in the free-fire area during the daylight, both before and after dark. Usually we left Can Tho about 4:00 PM and flew to Rach Soi to refuel, then spent several hours looking for bad guys before dark. Once darkness arrived we would go back to Rach Soi to refuel, rearm and standby on call. Then in the morning, if we had not been too busy during the night, we would return to the free-fire area to again search for targets of opportunity before returning to Can Tho about mid-morning.

Those of us who flew the mission often found we had trouble sleeping during the hot, muggy daylight hours at Can Tho, and unable to sleep at night while on call at Rach Soi. I was perpetually tired when assigned to this mission and tried to eke out a few minutes of sleep at night while on call, but the choices of sleeping accommodations were the rock hard floor of the Huey, or the runway outside. Neither proved acceptable, nor did they provide any escape from the bomber-sized mosquitoes that infested the place. The only successful defense against them was cigarette smoke, which they detested! But as soon as the smoke dispersed they again pounced on any exposed skin.

**Risky Business**

When I met him, Jack Orlo was a 21 year old warrant officer from Connecticut. Although he was four years my junior, he had graduated about seven months before I did and had been in Vietnam since August 1969, more than nine months before I arrived. I don't know if he had been with the 191st that whole time, but I suspect so.

Orlo was about 6 feet tall with a chunky build and reddish brown hair. His voice had a scratchy or gravelly quality to it, which was easily recognizable in person or over the radio. He liked coarse jokes at which he laughed loudly. He smoked and drank heavily and lived part of the time off-post with his Vietnamese girlfriend. Nothing about him seemed even remotely refined, and when he was at the controls of a Huey he was an excessive risk taker.

I was always uncomfortable being around Orlo because he was reckless and showed questionable judgment. I intensely disliked being his copilot because he got tremendous satisfaction being at low level with the lightship where he could flush out the VC quarry. The closer to the ground he got, the better he seemed to like it and the worse it was for his copilot who had to sit in the right seat with no control over what was going on.

On Thursday night, 11 June, we ventured into the free-fire area to look for bad guys. As usual, Orlo took us down to grass top level and began poking the skids and nose of the Huey into suspicious looking places. It was after dark, and the glare of the powerful xenon light danced on the grass that swayed violently under our rotor downwash. With Orlo flying, from my perch in the right seat it seemed like we were on a roller coaster and there was no way to get off. But finding no bad guys, he eventually gave up the game and we returned to Rach Soi to go on standby.

We were not called out that night, so early the next morning we were back at it in the free-fire zone looking for trouble, and this time we found some. Just as we were lifting away from a grassy hummock, a VC jumped up and began firing an AK-47 rifle at us. Orlo radioed that we were taking fire and the gunship rolled in to finish off the VC. As the firing ensued, I felt a sharp sting on my right shin, but could not look down to examine it since I was trying to be ready in case Orlo might be wounded and I would need to grab the controls.

We pulled up and away from the firing as the gunship engaged, and I looked down to see why my shin was burning. There was no bleeding to be seen and nothing was obviously wrong. But on closer examination I found the bootlace at the top outside eyelet of my boot to be broken. Holes in the chin bubble and the right door panel showed where a bullet had gone through the cockpit and bounced off the boot eyelet breaking the lace in the process.

When I told him what had happened, Orlo was greatly amused and seemed to think it indicated an important milestone in my experience. I failed to see the humor or the symbolism and I tried not to fly with him again. One by one I began eliminating the pilots in the company from my list of who to fly with, primarily on the basis of their judgment. At this point Orlo was at the top of the list, and he subsequently lost that position to only one other pilot who arrived in the 191st a few months later.

From a letter dated Sunday, 14 June 1970:

> ***On Sunday June 14th Rick Mahan and I flew ash & trash to Hotel 3 in Saigon to drop a couple of people off who were going home to the States, and flew three guys back to CTA. After lunch we flew to Chau Doc and then into Cambodia where we picked up prisoners who we flew back to Chau Doc. Then back to Cambodia again where we delivered mail and beer to outposts the rest of the day.***
>
> ***The mail and beer went to outposts where there was only one GI and about 50 ARVNs. Those poor guys lead a miserable life! Landed at one outpost***

*and the guy (a 1st Lt.) came running out in sandals and fatigues, looked like he hadn't seen a razor in a month. He just about jumped into the cockpit he was so glad to see us. All he lives in is a 2 man pup tent - no running water, etc. Food is all C rations or whatever he gets off the natives.*

*Stuck a cold beer in one hand and a letter in the other and he just grinned from ear to ear. He couldn't thank us enough. In fact, after we left, he came on the radio and thanked us again. God, I'm glad I'm not a grunt.*

From a letter dated Tuesday, 16 June 1970:

*Sure was tired when I got back yesterday. The mission ran much longer than it's supposed to. There were lots of VC running around south of Rach Gia and they wanted us to stick around and shoot up as much as we could. Managed to get one KBA (kill by air), 73 sampans and 22 hootches.*

*The area where we work down there is infested with VC and the officer in charge down there says it's a small part of North Vietnam. All we can do is go down there and shoot up as much stuff as possible so that Charlie has to rebuild it all the time. This keeps him busy so that he can't attack any of the frlendlies in the area.*

*When I ate last night it had been 24 hours between meals, and when I got back to bed it had been 48 hrs. since I got some sleep. We shouldn't have stayed out there that late. All we're supposed to do is fly a last light (at sunset) and first light (at sunrise) mission. Then stay on standby all night.*

*But, there was just too much to do and we couldn't leave.*

From a letter dated Saturday, 20 June 1970 referring to the Nighthawk mission:

*Last week an NVA Lt. Colonel turned himself in. They asked him why he did and he said that they work all day building hootches and bunkers, and at night the gunships come and destroy everything they build, plus keep them awake all night. Said he just couldn't take it anymore.*

**Air Medal**

Awards and decorations have historically been an important feature of military service almost as far back as the founding of the country. These recognitions generally fall into two classifications. The first is awards granted for being in a certain place at a certain time, like the Vietnam Service Medal which was awarded because one was assigned to the Vietnam theater of operations during the war. The second class is awarded for specific acts, sometimes for valor.

The Air Medal, created in 1942 by President Roosevelt, could be used for either classification and had the primary purpose in the Vietnam War of recognizing any air crew member who achieved 25 hours of combat flight time, or had a single act of merit or heroism.

By 14 June I had reached the first 25 hour threshold. I would eventually wind up my Vietnam service with 23 Air Medals, most of which were for accumulated flight time, but several of which were for other reasons explained later.

**Communications**

From a letter dated Saturday, 20 June 1970:

> *To tell the truth I'm getting very disgusted with the way things are being run in the company. I've never worked in an organization before where people cannot get along with one another. There seems to be no communication between the platoon leaders, between maintenance and anyone else, between the motor pool and anyone else, and especially between the CO and anyone else.*

Although there were many electrical devices available at CTA, like TV and radio, the one necessity that was missing was a telephone to call home. In order to do that we had to go to the MARS station (an abbreviation for the Military Auxiliary Radio System). MARS was comprised of licensed amateur radio operators who volunteered the use of their equipment and time to connect people in Vietnam to their friends and relatives back home.

The Can Tho MARS station was located in a small building on the north side of the airfield where one could sign up to get radio airtime to make a call home. The call itself was awkward and lacked privacy because there were radio operators between the caller and the listener monitoring the conversation. When one party in the call was finished talking, he would have to say "over" so the radio operators would know when to throw the talk switch to the other party, who would respond and then say "over"; and the first party could talk again. Even though it was awkward and not private, it was almost all we had for live communication with "the world" (which is how GIs referred to the U.S.).

The most prominent means of communicating with people at home was written correspondence, or "snail mail" as we say today. It was slow, but it was private and reliable, and news from home was always something to look forward to. I usually got a letter from my wife every few days. I was less reliable in return, partly because daily combat was not easy to talk about. Never-the-less I did manage to write over 120 letters to her, and more to parents and siblings.

For general news there was the *Stars and Stripes* newspaper, a daily that was published somewhere in the Pacific containing U.S. and world news, sports news, classified ads, cartoons and almost anything else one could find in a standard newspaper.

**Pleiku (*play-coo*)**

As a result of the retrograde of five Hueys and the damage done by the VC mortars on 1 June, the company was hurting for flyable aircraft and had not been able to fly CAs for a couple of weeks. So I wasn't surprised when I went to the O Club for coffee at 10 AM on 20 June and ran into CPT McConkey who said he was looking for a couple of pilots to fly to an airfield near Pleiku, called Camp Holloway, to pick up a Huey for the 1st Platoon. It sounded like an interesting mission to me, so I volunteered to go (I could tell by the way he talked that he was going to send me anyway). An hour later CW2 Bill Cotton, a Boomerang pilot, two other pilots and I were in a Huey headed for II Corps, the Military designation of the area encompassing Pleiku, Camp Holloway and the Central Highlands of Vietnam.

Pleiku, about 600 miles north of Can Tho, was (and still is) the capitol of Gia Lai Province. In 1970 it was a strategically important town that straddled the junction of National Roads 19 and 14, thereby controlling access to important cities in the Central Highlands, as well as Cambodia. Camp Holloway, near Pleiku, was smaller than CTA and was home to three Huey companies: a Chinook company, a heavy lift helicopter company of CH-54s, known to us as Skycranes, and an aircraft reconnaissance company of O-1 Bird Dogs. (It was a VC attack on Holloway in 1965 that provoked LBJ to retaliate by beginning the bombing of North Vietnam.)

We left CTA in a Huey and made it just past Saigon where we ran into severe weather. Finding that several of our navigation instruments didn't work and being short of good maps for navigation, we returned to Can Tho to figure out another plan to reach Pleiku. About 3:30 that afternoon, Bill and I, along with a crew chief and a technical inspector, got a lift with an Air Force plane from Binh Thuy AFB to Tan Son Nhut AFB in Saigon, and there we waited three hours for another flight. A torrential downpour that afternoon was clear evidence that monsoon season was fully upon us.

About 7:30 that night, we hopped another Air Force flight, landing about an hour and a half later at Cam Ranh Air Base, part of the gigantic logistics facility at Cam Ranh Bay built by the U.S. in 1965. We had no overnight accommodations arranged, so we slept wherever we could find a dry, flat spot in or near the terminal. The first spot I picked was on a set of bleachers, but it was impossible to get comfortable on a six-inch wide plank. While I was there a cute, black puppy, shivering from the cold damp air, came up and lay down beside me obviously looking for warmth and safety. I picked him up and snuggled him inside my shirt, but after a few minutes he got up and wandered off. By then the bleachers had lost their appeal as a bed, so I shifted to a large piece of sheet steel outside the terminal where I slept fitfully until first light.

By 8:30 on the morning of the 21st we had left Cam Ranh on another Air Force flight, arriving at Camp Holloway after 90 minutes in the air. The contrast from Can Tho was immediately noticeable upon leaving the plane: the air was cooler, cleaner and drier; the topography was rolling hills with mountains in the background, not flat like most of the delta; and there were no palm trees in sight. For "lowlanders" like us this environment was a huge change from what we were used to.

We hitched a ride to the Army compound and looked over the D model Huey we were to take to Can Tho. Cotton and I left the crew chief and tech inspector to examine the Huey's condition and oversee the finishing touches before the Huey was signed over to the 191st, and went to the pilots' lounge to wait. As the day wore on it became obvious that we would not be able to leave that day, and later on we were fixed up with beds for the night where we enjoyed a good night's rest in the cool temperatures of the Central Highlands.

On Monday, 22 June, we flew out of Holloway following Route 19 eastward above rugged mountain terrain to Qui Nhon (*kwin-yon*) on the coast of the South China Sea. There we turned south and followed the coast line to Tuy Hoa (*too-ee wah)* Air Force

Base where we landed to refuel. The coast line we followed was dotted with sandy beaches and in many places mountains ran right down to the sea, creating an unbelievably beautiful sight from 1,500 feet in the air.

We left Tuy Hoa and headed south to Ninh Hoa (*nin wah*) where we were cut off by bad weather, so we decided to return to Tuy Hoa to overnight. The room we got that night was shared with a couple of ARVNs, and it resembled a motel room in the States with its A/C, carpet, refrigerator and comfortable beds. We went to the almost-brand-new O Club that overlooked the beach for dinner that night, and had a nice conversation with a CW2 fixed wing pilot stationed at Phu Hiep (*foo hep*), just south of Tuy Hoa. With over 4,000 hours of flight time, he was only 18 months from retirement and in the conversation he told us the war activity in the area had slacked off because of the invasion of Cambodia, the same as we had experienced in the delta.

We hit the rack at 10:30, and after another good night's sleep took off from Tuy Hoa at 11 AM on the 23rd and flew to Phan Thiet (*fan thee-ett*) where we refueled. From there we flew to Vung Tau (*vung tah-oo)* where we ate lunch and picked up three GIs, then headed for home plate at CTA where we arrived in good condition at 5 PM. It had been a long trip of four days and 1,400 miles, but an interesting one. I got to know Cotton a little better, saw some beautiful country, broke the monotony of day-to-day operations, and brought home a badly-needed Huey.

That same day, 23 June, Al and a bunch of other guys flew to Qui Nhon to pick up three brand new H model Hueys, and with their return the 191st soon would be back in the CA business.

**Rach Soi Ambush**

The state of Texas is a very special place with a very special history, and has produced some very special people. One of them was CW2 Rick Mahan who had attended Sam Houston State University in Huntsville, Texas. Rick was a pilot in the 1st Platoon and quite the opposite of Jack Orlo in demeanor and risk orientation.

Mahan was about 5’8” tall with a medium build, sandy colored hair and a pitted but friendly face. He was even tempered and pleasant to be around, and often spent time at our hootch in intellectual conversation. Rick carried this manner into the cockpit and I thoroughly enjoyed flying with him because his style fostered confidence and cooperation. We not only enjoyed each other’s company, we worked well together.

On Wednesday, 24 June we were paired up to fly the C&C for the Rach Gia night mission, Rick being the PIC and I his copilot. By that time I had about 70 hours of combat flight time and was feeling more comfortable with the people I flew with. My confidence handling the aircraft had grown considerably and I was able to navigate effectively on my own, finding the way from Can Tho to Rach Gia, Rach Soi and around the free-fire areas with relative ease. By that time some of the shock of realizing what life was like in Vietnam was wearing off and daily routines were starting to develop. But I didn’t know just how non-routine this day would turn out to be.

We took off from Can Tho about 4 PM and arrived at Rach Soi airstrip 40 minutes later where we refueled and immediately went on standby. The next hour or so was spent eating a C ration dinner and smoking cigarettes until we received the map coordinates for that day's free-fire area. We cranked up our Huey and took off together with the lightship and gunship and headed out to hunt for bad guys, but there was little action to be found. After almost 2 hours of searching for VC and blowing up whatever looked like enemy bunkers, it was dark, the gunship needed to rearm and we all needed fuel, so we headed back to Rach Soi airstrip.

Reaching the strip about 10:30, the gunship landed first to refuel and rearm, followed by the lightship. We landed last, but took the forward-most refueling point about 50 feet ahead of the lightship and 100 feet ahead of the gunship. It was customary not to shut the Hueys down during refueling so they remained running at idle speed.

The refueling pads were actually PSP sheets about 12 feet square, lined up in a row just off the PSP runway on the north side. Like all landing strips in the delta, Rach Soi had long ponds of water lining both sides of the runway for its entire length, except where the refueling points were. The entire field was enclosed with concertina wire and ARVN guards were scattered around the perimeter to ward off VC who might try to get access to the airstrip.

By 10:30 it was pitch dark and there was scant lighting around the airfield. The only light we had to work with was from inside the Huey, or from the belly landing light under and behind the chin bubble. The shadows created by the crew members passing in front of the lights, along with a stiff wind that had come up from the west, gave the airfield a spooky look and feel, as if Halloween had come early.

After several minutes the gunship lifted off its refueling point and crossed the runway to the rearming point (a couple of Conex containers) where the crew began loading rockets and machine gun ammunition. Our crew chief finished refueling and climbed aboard, taking his seat in the transmission well opposite the door gunner. Mahan smoothly rolled on throttle and the main rotor blades gained speed quickly. As the rotor tachometer neared maximum speed of 6600 RPM, we heard a dull boom and a flash of light briefly lit up the area.

We were momentarily confused about what happened, but our crew chief said the lightship directly behind us had taken a hit from something fired from across the runway. As those words echoed in our minds the reality of what was happening started to sink in. We were being ambushed on what was supposed to be a secure airfield, at night, in the middle of nowhere.

Then we heard another round explode, this time under the chin bubble of the wounded Huey behind us. Shrapnel from the rocket round splattered up into the cockpit wounding both pilots in the legs making a takeoff impossible for them. They were sitting ducks.

All at once Mahan jerked our Huey to a hover. There was enough light in the dark cockpit to see that his easy-going manner had changed and he was all business. As he dipped the nose forward for takeoff we heard a third explosion near the lightship and felt a concussion of air pass through the cockpit. Mahan was pulling all the power out of the Huey he could get and as we reached about 50 feet of altitude a fourth round hit the PSP and detonated directly under our tail boom.

Immediately Mahan yelled, "Get on the controls with me; we don't have any hydraulics." When the fourth round exploded, it sent bits of shrapnel hurtling into our Huey and one of them severed the main hydraulic line. Although we had been trained in flight school how to handle a hydraulic failure, the truth is it was nearly impossible to fly a Huey without hydraulic assist.

I grabbed my set of controls and together we began to wrestle with what had become an un-flyable beast. The foot pedals, cyclic and collective seemed like they were welded in place, almost totally unmovable. Mahan yelled again to push the cyclic forward and lower the collective so we could land, but that seemed like a bad idea that would take us right back into harm's way and I thought we should continue the takeoff. As we struggled to move the controls even a tiny amount, I realized we would never be able to fly the chopper in this condition. It was much worse than what flight school instruction had prepared me for.

We were straining to push the collective down and decrease power, and to push the cyclic forward to level the Huey's nose from takeoff attitude, but nothing was moving and we were still climbing. Mahan started beating the collective with his hand and I was pushing down on it with all the strength I had, and all at once it moved ever so slightly. We kept pushing and beating on both the cyclic and collective, gaining a little with each blow until we began to level off. But as soon as this victory was achieved we realized we were in an uncontrolled descent with almost no way to stop it.

Meanwhile, when the Bounty Hunter pilots realized what was happening they dropped their rocket pods to lose weight so they could get into the air quickly, and as we dropped toward the runway they streaked by us taking off. It quickly became obvious we were going to hit the runway hard, but that was not our biggest concern. We still didn't know where the VC were, or what they were up to next.

Our Huey hit the PSP like a ton of bricks. I heard the sound of ammo boxes, C rations and other gear sliding across the cargo floor behind us, whacking against the backs of our seats as our heads snapped forward like rag dolls. The force from hitting the PSP was unexpectedly strong and we simultaneously grunted "umph" as the impact forced the air from our chests. Mahan rolled off the throttle and as the main rotor began to slow he yelled at the crew chief and door gunner to get out of the Huey and find some cover. It was obvious from his movements that he was panicky and confused, torn between taking time to shut the Huey down and running for his life.

The VC had fired from Rick's side of the aircraft and we both figured that's where another round would come from, if they had another round left. I figured that if a round was fired at us, Mahan was in the most vulnerable position because the firing was coming from his side, so I told him to get out and I would shut down the Huey. At this suggestion he didn't hesitate and was unbuckled and out the door instantly, while I found myself alone trying to fight back the terror that was trying to overwhelm me. Now, frantically, I mindlessly began pawing just about everything I could see hoping to hit enough switches in the process to shut the Huey down.

I eventually hit the right combination because the engine started to shut down. As the main rotor lost inertia and began slowing, the red cockpit lights stayed on, powered by the battery alone. I looked down at my lap searching for the quick release of my safety belt, which was buried somewhere under the edge of my chicken plate and tucked behind the handle of my holstered .38 revolver, stuck in the crotch between my thighs.

I was terrified, expecting to be blown to bits at any moment as I dug under the chicken plate for the belt release. I could hear myself whining and grunting as I found the release and pounded it with the butt of my hand. I glanced out the left cockpit door that Mahan had left open thinking maybe I could see if any VC were sneaking up on me. But no matter how hard I peered into the darkness I couldn't see anything.

I pulled off my flight helmet and could hear the welcome sound of the Bounty Hunters circling about 200 feet above the airstrip looking for the bad guys. It was reassuring to know they were up there, looking out for us Boomerangs as usual. I could also hear yelling coming from the direction of the lightship which was still at the refueling point a couple of hundred yards to the rear.

I went back to beating on the belt release, but it wouldn't budge and I remember thinking what a piece of junk it was if it wouldn't open easily in an emergency like this. After numerous jabs it finally gave way. I threw off the shoulder harness and grabbed the cockpit door handle next to me only to find that it had jammed when we hit the runway and wouldn't open. In the space of a few seconds terror turned to panic as I turned the handle and beat on the door over and over.

My imagination was in overdrive picturing various macabre scenes. I saw a VC round hitting the Huey in the fuel tank which then burst into flames, roasting me in the cockpit.

I turned and looked at Mahan's door standing wide open and decided that was the only way out. I crawled over the center console and across the left cockpit seat, falling out on the runway where I could hear Mahan and the crew calling me from a water filled ditch on the other side of the Huey.

Drawing my .38 I ran for it, jumping in feet first next to Mahan and the crew. The water was warm and chest high (and probably full of leeches I thought). Mahan had his .38 out too, but the rest of the crew was unarmed.

We scanned the area across the runway and saw nothing in the dark except for light reflecting faintly off the concertina wire at the outer edge of the airfield. Looking down the runway to the left we could see the lightship on the refueling pad with people moving around. It was clear that even though the Bounty Hunters were still circling we were far too exposed. Having no protection and realizing we had only a few .38 rounds between us, I began to recall the hand-to-hand combat training we'd had at Fort Benning, expecting it might be needed soon.

A minute ticked by and all of us seemed to be OK, so I asked Mahan if he knew the condition of the lightship crew; he said he had not heard anything. Since we had no way to communicate with the gunship and felt too vulnerable to just wait there in the water, I decided to make my way back to our Huey to get the emergency radio that was kept in a box under the copilot's seat.

By now the terror and panic had begun to subside, so I slithered out of the water and onto the PSP. The cold steel of the metal could be felt through my soggy flight suit as I low-crawled to the side of our Huey about 25 feet away, my eyes straining to pick out any VC in the darkness. I stood up quickly and jumped into the Huey through the right cargo door, then quickly reached over the center console and grabbed the red-painted ammo box under the copilot's seat. It clattered against the metal rods beneath the seat and banged against the side of the center console as it was yanked free.

I slid off the cargo floor and ran upright back to ditch and jumped in again. I dug out the radio and talked to the Bounty Hunters above us who said they only had machine guns and had dropped their rocket pods before taking off, so they couldn't provide rocket cover. I asked them to contact the Seabee unit in Rach Gia and tell them to send out their Seawolves for additional cover. I also asked if they could see VC anywhere, but they said no and advised us to stay put in the water until help arrived. Oh great, I thought, more time for the leeches to do their work.

A couple of minutes later, the wounded pilots and crew chief from the lightship came hobbling toward us and lay down next to the ditch where our crew started looking at their wounds. They told us the lightship was a mess and the door gunner, Tim "Porky" McCarthy, had been killed when the second round went off. McCarthy had been refueling the lightship and after the first B-40 rocket round hit he dropped the fuel hose and ran to the other side of the Huey to get in, the side from which the rounds were being fired. As he reached the other side a rocket hit the PSP at his feet, killing him immediately. They also described what happened when the round that hit them came in, shaking as they told their story.

As they described what had happened, a few sleepy ARVN soldiers, the ones who were at the guard posts supposedly securing the airfield, came straggling across the runway rubbing their eyes like they had just awakened. I wondered if they were actually the VC who had fired on us and were feigning sleepiness to cover up their deeds. Mahan and I kept watch of them as our crew started to attend to the other crew's wounds.

Our sad bird sat disabled on the runway 25 feet away, its red navigation lights that I had failed to turn off still burning, its rotor quiet. Several hundred yards down the strip the lightship too was still lit up, but it was in far worse shape.

The wind was still steady from the west and chilled us as we stood in the water. Gradually our shivering from fear turned into shivering from the wind, and it seemed ironic that in such a perpetually hot place where one longed for relief from the heat and humidity, one could actually be cold.

We waited there in the water for some minutes, wondering what would come next and trying to process what had just happened. For those of us experienced in the Nighthawk mission, this experience was unique. In just a few minutes our daily routine had morphed from the familiar to the unfamiliar, from something spooky to something disastrous and deadly. And I think we all realized this ambush could have been much worse.

A few minutes passed before the Seabees' gunships arrived and started circling the strip with the Bounty Hunters. One of the Seawolves landed and the wounded crew members got on board, then evacuated to 3rd Surgical Hospital at Binh Thuy.

My boots and pants were covered with mud and I couldn't wait to get my shirt and pants off to check for leeches. We stripped to our skivvies and carefully searched, and burned some leeches off with cigarettes as reinforcements began arriving! Five

Boomerang crews came in and landed and I was never happier to see people who I knew and could trust to be on our side. It was just like in the old western movies when the cavalry shows up.

The rest of the night has been lost in memory, except for one final detail. One of the Boomerang pilots who came to our rescue, a warrant officer named Jensen, quietly approached me and asked if I would fly his Huey back to Can Tho since our bird couldn't fly. I figured I was in no condition to fly anything, so I asked him why. He told me that, when the news hit the O Club that we had been shot down, all the Boomerang pilots emptied the place and headed for the flight line. When he got to his Huey the crew was there, but he had no copilot so he flew out alone. He thought it would be a good idea if I flew back with him because he thought he was too drunk to fly!

I was amazed as he told this story, which brought a mix of admiration for his dedication and commitment, and disgust that he would jeopardize himself, his crew and the aircraft to fly alone in that condition. I asked him why he would do something that risky and he replied emphatically, "Because I was worried about you!" I was humbled and grateful that the Boomerangs were concerned to that degree for the welfare of each other.

By midnight of 24 June, with more than 10 months left on my tour, the odds of survival looked bad. In the six weeks I had been in country, five members of the company had been killed and several more wounded. At that rate nearly a quarter of the company would perish before I could go home, and I didn't want to be one of them. The question of what to do to survive seemed beyond answering when battle death was so arbitrary.

The ambush itself lasted only a couple of minutes, but among other things it produced the intense emotional reaction the enemy had no doubt intended. In a split second my mental state had gone from calm to terrified, and then over the next couple of hours it dribbled off to a more controllable level of anxiety. One exposure to terror like that might eventually dissipate on its own, but when it was piled on top of other similar experiences the consequences became deeper and more profound. That night, ordinary alertness had been replaced by paranoia.

About three days later, there was a memorial service in the company for Porky McCarthy. A symbolic M-16 was mounted upside down on a table with a helmet on top and combat boots in front of the M-16. I remember thinking that those spit-shined boots were not representative of the person whose memory was being honored, because Porky never shined his boots. Most of the company pilots were present and many of the enlisted men. The company commander spoke a few words, but they seemed inadequate. It was less than enough for a man who had died doing the duty his country demanded in a foreign land.

The ambush at Rach Soi resulted in a Bronze Star for valor. It came less than two months into my tour, for action that didn't seem valorous to me. I had done what the situation called for, and I suppose that's what many people say who've been decorated for valor. In moments of great danger you do what you need to, and sometimes other people see it as heroic.

**Chieu Hoi (*chew hoy*)**

That same night before Mahan and I were shot down at Rach Soi, we dropped propaganda leaflets over several small villages in the U Minh (*yoo min*) Forest south of Rach Gia. The Chieu

Hoi Program (loosely translated as "Open Arms") was an initiative to encourage defection by the Viet Cong and their supporters, and to support the South Vietnamese government. From Wikipedia:

> Defection was urged by means of a propaganda campaign, usually leaflets delivered by artillery shell or dropped over enemy-controlled areas by aircraft, or messages broadcast over areas of South Vietnam. A number of incentives were offered to those who chose to cooperate. Overall the Chieu Hoi program was considered to be successful. Those who surrendered were known as "Hoi Chanh", and were often integrated into allied units as Kit Carson Scouts, operating in the same area where they had defected. Many made great contributions to the effectiveness of U.S. units, and often distinguished themselves, earning decorations as high as the Silver Star. The program was relatively inexpensive, and removed over 100,000 combatants from the field.
>
> Most early Kit Carson Scouts had defected as Hoi Chanh Vien because they suffered either from malaria or grave wounds beyond what could be medically treated with the rudimentary medical care available on the Viet Cong/NVA side. Those Chieu Hoi who volunteered for selection and training as Kit Carson Scouts had, during their service with the enemy, little or no contact with anyone speaking English. Few had any knowledge at all of the English language, creating a communications challenge as they were deployed with American units. A further complication was that almost all Hoi Chanh Vien had a distrust of Vietnamese soldiers and interpreters because of the degree to which friendly forces had been infiltrated by enemy agents.

Not wanting to give my wife anything to worry about, I only mentioned dropping the Chieu Hoi leaflets in my next letter to her on 26 June. I did not mention the ambush at Rach Soi.

--- Ω ---

The war in Vietnam was like a game of chess being played by the proponents of competing political ideologies: communism vs. capitalism. Or, maybe it was a line of deadly friction where the power of the "haves" in western societies rubbed up against the desires of the "have-nots" in eastern societies. In 1970 I convinced myself that it was the former, but experience has taught me that it was likely the latter. In either case, the U.S. troops who served in Vietnam were compelled by law to serve in a war that was conceived and executed by the U.S. President, not by the U.S. Congress, which seemed to be a clear violation of the U.S. Constitution.

I had made a conscious choice to comply with the orders of President Lyndon Johnson (LBJ) and his puppets, and because of this conscious decision I share blame for the war as a mindless pawn (in lockstep with the old Army adage, "Ours is not to question why; ours is but to do, and die"). But the innocents in Vietnam and the bordering countries of Laos and Cambodia, whose meager existence required almost all their daily attention, suffered beyond imagination as blameless victims. Helplessly caught at the line of friction, they were swept up and destroyed in a tornadic blood bath that destroyed anything and everything even close to its path. Millions suffered; millions died.

By the end of June the invasion of Cambodia was coming to an end. There was no way for us to know if easing up the pressure the "incursion" had put on the NVA and VC would result in stepped up combat activity in the delta or not. But the paranoia that resulted from the ambush at Rach Soi on the 24$^{th}$ had prepared me for just about any eventuality, except the humiliation that comes from an act of sheer stupidity, as demonstrated five days later.

## The Red Box

Every Huey crew had a red box that contained essential survival equipment like a strobe light marker and a small confidential paper booklet called a SOI (Signal Operating Instructions). The SOI was a type of combat order issued for the control and coordination of communications within a command, like the 13th Battalion. The book included up-to-date information covering radio call signs and frequencies, a telephone directory, and code words for encryption.

The box also included a small hand-held radio, two flares, a large plastic map of the Delta, and some "blood chits," paper documents that were used by downed air crews to reach safety. The blood chit that we had translated the following phrase into Vietnamese and 12 other languages:

> *I am a citizen of the United States of America. I do not speak your language. Misfortune forces me to seek your assistance in obtaining food, shelter and protection. Please take me to someone who will provide for my safety and see that I am returned to my people. My government will reward you.*

The reward for returning an American under this program was $10,000, an astronomical sum for almost any South Vietnamese.

The red box was actually a metal ammunition can that had been painted. It was air and water tight, and if a crew crashed it could be removed from the Huey and its contents used for survival purposes. It contained highly sensitive information that no one wanted in the enemy's possession, and we had been cautioned to safeguard it. By tradition, it was the copilot's duty to keep track of the red box and account for it at all times, but the final accountability actually rested with the pilot in charge, or aircraft commander (AC).

On the 29$^{th}$, I was copilot on a single-ship mission to Moc Hoa in support of the 5$^{th}$ Special Forces Group. Our morning flights delivering mail and other such activities were easy and uneventful, and about noon we knocked off to get some lunch at the Special Forces compound in downtown Moc Hoa.

We shut down our Huey in an open field that was being baked dry as a brick by the relentless sun. Several ARVN soldiers were standing by and agreed to keep an eye on it for us. We were picked up by a Vietnamese driver in a van who tore down the narrow, palm-lined dirt road to Moc Hoa at 80 mph, passing buffalo carts and pedestrians like they weren't even there and leaving them in a cloud of dust. He stopped at a low-walled compound about 100 by 100 feet square that enclosed a number of low brown buildings, including a small dining hall where we ate an unmemorable lunch, and then rode back to our Huey with the hell-bent-for-leather van driver.

As soon as I opened the copilot's door I noticed that the red box, usually stowed underneath the copilot's seat, was gone. My heart stopped. I glanced around the cockpit floor thinking it might have slid out of sight, but no. It was definitely missing. My heart stopped again!

I swallowed hard, and the AC heard, "Oh, shit!" as the words burst across my lips. He wanted to know what was wrong. I looked around again and found our flight helmets, the crew chief's and door gunner's M-60s, the ammunition and grenades.

Everything except the red box was still there. I breathed a partial sigh of relief, then told him what was up. He was not happy.

We asked some of the local guards where the box was, but they just shrugged and acted like they didn't know what we were talking about. Searching for it seemed futile as there was no obvious place to start. So, even though the AC and I knew this incident could result in a compromise of the radio communications in the battalion, or an accusation of negligence, or something worse, there was really nothing to be done but to report the loss when we returned to Can Tho. And take our lumps.

Filling out our after-action report when we got back to the Boomerang operations office was agonizing since it included the embarrassing admission of such a basic security error. Then too, there was the anxiety of a possible official reprimand. But as it turned out there were no consequences other than a look of disgust and a muttering of "Dumb f___s" from the company operations officer.

We ducked any serious consequences at the time, but I have never escaped the self-condemnation that resulted from this incident. To this day I am humiliated to admit such a basic failure of responsibility.

## Chapter 3 - JULY 1970

Almost two months down; more than ten to go.

| *Mama Told Me Not to Come* - Three Dog Night |
|---|

I referred previously to a "chicken plate" which was the nickname for the Kevlar body armor worn by Huey crew members across the chest and stomach. Chicken plates were about three-quarters of an inch thick and slid into a fabric vest that was worn around the upper body with the plate in front. Velcro tabs fastened the vest.

Crew chiefs and door gunners, who were not strapped in seats and had to be on their feet to fire their M-60s, wore them with the vests. However, some pilots found the vests to be bulky and even dangerous in some circumstances, so they pulled the plates out of the vests, which were discarded. The plates, which came in several sizes from small to extra large, were placed over the chest and abdomen and rested on the pilot's lap, fastened beneath his shoulder harness. (I wore mine with the vest, as did all the pilots in the 191st.)

Each pilot had his own personal chicken plate, selected based on how well it fit over the chest and stomach area. Usually the larger the pilot, the larger the plate, but not always. There were exceptions, and CW2 Gordon Clayton was one of them.

Clayton was only about 5'6" tall, but he had one of the largest chicken plates in the company. When he was strapped in a Huey seat, the plate was so large it covered the lower part of his face almost to his nose. I flew numerous missions with Gordon and recall many times being able to see only his eyes in the space between the top of the chicken plate and the bottom of his flight helmet. Hunkered down behind his plate, peering through that slit with sunglasses on, he looked like Darth Vader from the *Star Wars* movies.

Gordon was small but wiry, with thinning black hair and a thin black moustache. He was good natured, soft spoken, and a competent and reliable pilot. When things got challenging in the cockpit, he never raised his voice or used foul language. When he laughed he snickered, and his shoulders would bounce up and down. Mild mannered, gentlemanly and outgoing, everyone liked him, including the enlisted men.

**Puff**

The recent disaster with Rick Mahan at Rach Soi was no longer at the top of my mind. It had become buried under the incessant missions that piled one on top of the other, day after day. There was no time to really analyze the experience, and maybe that was good. I didn't want to dwell on it, which would only reinforce that our situation was untenable and confirm that there was a choice between only two unacceptable alternatives: desertion or death.

It was early July and Clayton and I prepared to fly C&C for the Nighthawk mission. We had flown it together several times recently and were getting used to each other, he as the AC and I as the copilot. We left CTA and reached Rach Soi about 4:00 PM, and after refueling and receiving the map coordinates we flew south to the free-fire area. The next few hours were mostly uneventful, but the gunship pilots used up enough of their ordnance that they had to return to rearm several times while we stayed on station. The weather was clear with a broken layer of clouds several thousand feet above us.

It was fairly quiet until dark, but then the situation changed completely. We were several klicks (kilometers) south of Rach Gia waiting for the gunship to return when we started receiving calls for help from the operations center. Several friendly outposts were under attack at the same time and needed our firepower to suppress the enemy. Clayton received the info and directed the lightship to the outpost locations a little farther south of us, and radioed the Bounty Hunters to join us there when they were rearmed.

When we arrived at the scene we could see small arms tracer fire coming from tree lines that were barely visible in the moonless night. The lightship was putting machine gun rounds on the enemy locations as fast as possible, but when one location stopped firing another one a distance away would start up. The Bounty Hunters arrived and expended their rockets quickly on the growing number of targets, then had to return again to rearm. A round trip for them was about 30 minutes, so during that time only the lightship was available to put down covering fire, and that wasn't enough.

It went on this way until we and the lightship had to refuel, and we left for Rach Soi. In our absence the situation on the ground worsened and after returning it was clear we were not keeping up with the growing number of attacks. We started getting urgent calls directly from the outpost commanders or their American advisers who were with them, asking us to get more firepower on their positions to relieve the pressure of the attacks. We radioed the Seawolves in Rach Gia who sent out their gunships loaded to the hilt, but their ordnance was soon expended. They left the scene and did not return.

It was now approaching midnight and we had been in our seats without a break for almost eight hours. Fatigue was growing and the muscles in our legs were beginning to cramp. I craved a Marlboro and a drink, and would have been happy with a five minute stretch, but no relief was in sight. I wanted to complain and I'm sure Clayton did too, but we knew that wouldn't help so we bit our tongues and didn't talk about it.

The Bounty Hunters left and returned again, but each time they would leave to rearm or refuel the attacks on the outposts would revive. So, Clayton got on the horn (the radio) and called the operations center for reinforcements. As it turned out practically no one was available to help us in the entire delta area that night, except for a pair of Air Force OV-10 Broncos with the call sign "Black Pony."

In 1970, the Bronco was a fairly new plane in the military inventory. Designed for close ground support, it was only 44 feet long with a 40 foot wingspan and could take off on a very short runway. It flew fast and packed a big punch, and within a few minutes we welcomed the Broncos into our airspace.

Clayton told the Bounty Hunters to vacate their low level position to make room for the Black Ponies, and then told them where to place their fire. They were tremendously accurate and effective in suppressing enemy return fire, but they were only there about 10 minutes before their ammunition was expended. They too departed and didn't return.

By then another hour had passed and we were facing more fatigue and a stubborn enemy. It seemed like there were attacks on outposts as far as we could see with no let up in sight. I began hoping that we could hold out till daylight so the outposts would continue to have whatever cover we could provide, but the situation looked bleak. We couldn't even get reinforcement from the 191st because they were scheduled to start combat assaults in just a few hours and had no one to send.

We continued to shuttle back and forth to Rach Soi for fuel and ordnance, one at a time. Another hour passed while the tracer rounds from the tree lines continued their onslaught. We had no way of knowing what was going on at ground level, and really didn't want to know. The situation was a reminder of the advantage a pilot had by being at 1,500 feet instead of at ground level with the infantry, a place I would have been had I not attended flight school. I was mighty grateful to be above the fray, even if my legs were literally numb to the point of not being able to feel the foot pedals any longer.

The situation persisted this way until a voice came over the radio that said, "You need some help down there?" Clayton and I looked at each other for a few seconds in mild surprise, struggling to overcome the mental lethargy brought on by too many hours in the cockpit without a break. He keyed the pilots' intercom and asked me if I had heard that, and I said I had. Then he keyed the FM radio and asked the lightship and Bounty Hunter pilots if they had asked if we needed help, and they said no.

We looked at each other again and realized that someone else might be out there that we didn't know about. There was only one thing left to do: Clayton flipped the radio transmitting switch to the UHF guard channel and said cautiously, "Who's this?" The answer came back immediately, "This is Puff." We peered out into the darkness looking for another aircraft, but all we could see were the distant lights of Rach Gia along the coast, the tracers from the VC rifle fire, and our own Hueys working the area as best they could.

Clayton keyed the radio again and said, "Puff who?" Again the response came back immediately, "Puff the Magic Dragon." At that point we knew what we were dealing with as we both knew what Puff the Magic Dragon was, even though we had never seen one. But we could not see where he was until we looked almost directly above us. Circling silently at about 4,000 feet, partially obscured by the intermittent cloud cover, was the bulky body of the World War II era AC-47 that was Puff. It was too dark to make out the color of the plane, but we could see its navigation lights easily once we knew where to look.

Clayton affirmed that we needed help and explained where the enemy locations were in the tree lines. At Puff's request our three Hueys flew clear of the area and Puff began to lay down fire with its electric Gatling guns at more than 6,000 rounds a minute, directly on the sources of the VC small arms fire in the tree lines.

Every sixth round in Puff's gunfire was a tracer round (a round that gave off light as it traveled), but the firing rate was so high that the streams of tracers looked like they were solid lines of light. Puff's guns would fire a short burst for several seconds, then adjust a little and fire another short burst. The firing was accurate and effective and the enemy firing slowed, then stopped altogether. It was an awesome display of firepower that completely overwhelmed the opposition. Clayton and I were practically ecstatic at the effect it had on the situation, and the outpost commanders reported that enemy fire had abruptly and completely stopped when Puff's work started. Puff had stepped in when we badly needed help and the action looked like it was over.

By now it was about 3:00 AM and we were physically exhausted from almost eleven hours of non-stop flying. Needing to move on, I suppose to more fertile opportunities, the anonymous Puff pilot wished us good luck and left the area. We felt naked and exposed, and within a few minutes the enemy fire had resumed. It seemed like we were right back where we started, but at least we were an hour closer to dawn.

At some point during the night, both Clayton and I crawled out of the Huey when we were refueling and stretched our legs for a couple of minutes. Then we dragged through the next couple of hours, shuttling to Rach Soi and back until light started to show on the eastern horizon. The enemy firing began to subside, first here and then there, until it stopped completely. By the time it was fully light the attacks had ended.

We flew back to Rach Soi to refuel and rearm and crammed down some cold C rations to quiet the starvation that gnawed at our guts. We shut down the Huey and went on standby for the first time since leaving Can Tho Airfield the day before. Satisfied that there was no more we could have done the night before, we hoped our presence had made a life-saving difference to friendly troops on the ground.

Within an hour we were released by the operations center at Rach Gia and returned to Can Tho. I was too tired to wash off the dirt and sweat from the night before, but in spite of it my Army cot never felt better.

Today the details of that night are vivid memories of Seawolves, Broncos, Puff's tracers, enemy fire from tree lines, physical exhaustion, and an always-present anxiety, all combining to stand out as a unique experience. It was far away from home then, and is far away in time now.

**Spooky**

The average elevation of the Mekong Delta was below sea level, which made it difficult to find solid ground to build anything on. But that was not the only water we had to contend with, for by July we were in monsoon season when it rained every day. Adding in the constant high humidity, it seemed like we were surrounded by water all the time.

Water complicates combat flight operations. Often during the monsoon season we would land to a rice paddy full of water, or some bedraggled area where tall grass would grow out of muddy water, places that would be completely dry during the dry season. On more than one occasion we had to land to a hover in a watery LZ and the rotor downwash would lick up the dirty ground water, throwing it on the windshield and making it impossible to see to the front. Removing the water with the wipers on the Huey was a no-no because grit on the plexiglass windshield would scratch the windows making them permanently opaque. So when the windshield became obscured by dirty water, the only way to see where we were going was for the pilot to stick his head out the side door window. It was an awkward way to fly, but it worked, and often the monsoon rains would wash the windshield clear in a few minutes.

Rick Mahan and I were paired to fly Nighthawk on 4 July, but the preflight inspection turned up numerous maintenance problems and we didn't get off the ground in Can Tho until 7:00 PM. Two-thirds of the way to Rach Gia we found ourselves surrounded by thunderstorms and had to return to CTA. The weather at home base was 0-0 (0 feet of visibility and 0 feet of ceiling due to heavy rain, a no-fly condition), so we stayed at CTA waiting for the weather to improve. The operations center at Rach Gia was notified of the weather problem and they agreed we should stay on standby at CTA.

The rain turned torrential, pounding down in buckets without letup. We waited in the operations shack until 11 PM, then hit the sack expecting we would see no action that night. About 3:00 AM, the battalion operations center called and said that several outposts were being attacked north of Vinh Long in a place named My Phuoc Tay *(me foo tay)* in Tien Giang *(tin zhee-ang)* Province, and they needed our help. It was still pouring, but the call was officially dubbed a tactical emergency which allowed us to take off even under 0-0 conditions (tactical IFR).

We three Nighthawk crews headed to the flight line and crawled into our Hueys with water running out of our boots. We cranked up and hovered to the runway, waiting for clearance from the tower to take off. The rain was incredibly heavy in the pitch black conditions. Sheets of rain blew across the windshield and the rotors kicked up water from the runway giving us only a few feet of visibility. The runway itself could be seen below us, and maybe 30 feet to the front in the glare of the Huey's belly landing light, just enough to see the tail of the Huey to our front. The red rotating beacons above the pilots' cabins turned the raindrops red each time the light beam flashed by. It was at once disturbing and beautiful.

The AC in the lead Huey had a plan to deal with the weather: he radioed specific instructions to us saying the lead Huey would take off and climb at 500 feet per minute on a compass heading of 080 degrees at 80 knots; the second Huey should take off 30 seconds later and climb at 450 feet per minute on a heading of 075 degrees at 75 knots; and the last Huey should take off 30 seconds after that and climb at 400 feet per minute on a heading of 085 degrees at 70 knots. The idea was to maintain separation to avoid a midair collision until we broke through the tops of the clouds, which the weather report told us would be several thousand feet up. It sounded workable.

The tower cleared us for takeoff and the lead Huey dipped its nose and disappeared immediately into the gloom. Ten seconds later, the second bird dipped and took off, and also disappeared instantly. Then it was our turn. Mahan lowered the nose moving us forward until we reached translational lift, and then nailed our climb to 400 feet per minute on a heading of 085 degrees at 70 knots, as planned. The wind and rain kicked us about like a rubber duck in a bath tub as we climbed through the pitch black conditions. Maintaining the takeoff instructions was difficult due to the turbulence, and there was no sign of the lights from the first two Hueys ahead of us.

A minute passed, and another, then lead radioed to say he had not broken out of the clouds yet. Mahan and I were glued to the instruments trying to hold to the course we'd been given, but the airspeed indicator was jumping all around and the actual speed had to be interpolated every second. The wind, rain and gloom were incessant.

Four minutes passed, then five. Finally lead radioed that he had broken out of the clouds at around 2,500 feet. Soon the second Boomerang ship radioed to say they were out of the clouds too and it was a beautiful, clear night. And then we too broke out of the clouds as the rain abruptly ended along with the severe wind turbulence.

It was indeed a beautiful night up at altitude, as clear as if it had been washed. There was no moon, but there were hundreds of crystalline stars that stretched across the entire sky. The clouds beneath us blocked the lights on the ground, making the stars even more brilliant. The air was much colder above the storm, and to our left ahead of us we could see the flashing lights of the other Hueys that had turned north toward Vinh Long, their sharply defined forms silhouetted against the starry sky.

It was about 30 minutes until we arrived above the outposts that were being attacked by an estimated VC battalion (around 350 guerrilla soldiers). They were receiving a lot of fire, but as soon as we started to fire back the VC fire dropped off. About a half hour later a Dustoff (i.e. a medical evacuation helicopter) flew in to pick up nine wounded or dead ARVN soldiers and flew out unscratched.

But upstairs we were having problems flying C&C. A Vietnamese Air Force (VNAF) "Spooky" had shown up and we were having trouble communicating with the pilots. Spooky was the same plane that we Americans knew as Puff and had the same lethal equipment aboard. And no matter what we attempted to do we could not establish radio contact with the pilots. We spent the next 45 minutes dodging Spooky, and then went to Vinh Long AAF to refuel and rearm. By the time we were back on station over the outposts, the VNAF had disappeared and we spent another couple of hours there. By then the action had nearly ended, and by dawn we made our way back to CTA where I spent the next 24 hours running back and forth to the latrine with dysentery!

It had been a great learning experience and a gratifying mission. Boomerang lead's unique instructions for takeoff were a product of the needs of the situation, and like so many other things I was learning about warfare, flexibility and creativity were important to successful outcomes.

I was flattered by a second award for heroism for this mission, although I never knew who provided the nomination for the decoration. Again, I thought I was just doing my job, which required no exceptional courage.

At this point in my tour I had over 100 hours of combat flight time, much of which had been on the Nighthawk mission as copilot with Orlo, Clayton and Mahan. It was a mission I understood and felt somewhat proficient in. However, I'd had little experience in formation flying due to the lack of Hueys in the company, and the 191st had gone almost six weeks without flying any combat assaults. To this point I'd only flown CAs twice in the nine weeks since arriving at the 191st and still had a lot to learn, as the following story reveals.

**CW2 Denver Stiles**

Experience is a good teacher, so they say, but it's also an expensive one when the cost of gaining it is taken into account. Even so, gaining knowledge through experience is sometimes the only way it can be done.

On 10 July, I was flying copilot in only my third combat assault with a CW2 aircraft commander (AC) named Denver Stiles. Flying to the west out of Soc Trang, we were in the second position in the right leg of the V-shaped flight formation, and as we made our approach to the LZ landing instructions were given to the flight. Because the LZ was very small and surrounded by tall trees on three sides, exiting it would require turning our Hueys around 180 degrees and leaving on the same path we flew in on.

Stiles was letting me fly to build my experience, but he was obviously uncomfortable and his discomfort made me nervous. After consulting by radio with the rest of the flight, he gave me specific instructions: after the ARVN troops were off our choppers, we were to turn our Hueys around with the tails away from the rest of the flight to be sure we didn't collide with the tails of the other Hueys as they turned. I remember hearing his instructions clearly, but my attention was riveted on controlling the Huey during the descent to the LZ.

The 10 M-60s in our flight were spraying lead into the nearby tree lines during the bumpy descent. The smell of burned JP-4 fuel was everywhere as the Bounty Hunters' rockets hit the LZ which was erupting in geysers of grass and dirt. Our landing ended at a three foot hover with the rest of the flight, surrounded to the front and sides by lush, vibrantly green trees 50 or 60 feet high. The sun through the windshield blinded any view directly to the front and boiled our faces through the perspiration that poured down them. The LZ was dry and the grass beneath us was tall, nearly touching the belly of the Huey. It jumped and swayed erratically in the rotor downwash as other pilots in the flight reported taking hits from VC small arms fire.

The ARVN troops started jumping out, and after what seemed like minutes instead of seconds the crew chief said all the troops were off and clear of the tail boom. I started to make the turn for our departure, but instead of turning the tail to the right, to the outside of the flight, I turned it to the left, i.e. toward the flight. Stiles yelled at me to stop the turn and go the other way, but it was too late for that since I was halfway through the turn by then. Already high anticipating the possibility of enemy fire, my anxiety ratcheted up as I realized my mistake. My legs went rigid. I clutched the controls like a vise. I nearly froze.

In that moment of near panic, my thoughts quite amazingly became detached from the circumstances as they tried to lay out a strategy for overcoming a dangerous and potentially deadly maneuver. I was aware of two voices in my head at the same time, one telling me I was going to die if I didn't get a handle on the situation, and the other simply pleading for safety.

Stiles was flustered, but he recovered his composure as we took off with the flight, leaving the VC fire behind. I felt like an idiot, and he must have realized it because the only thing he said was I needed to be careful to follow all the instructions that we were given to avoid making a critical mistake that could result in destroyed helicopters and lost lives. While I was grateful for the gentleness of his admonishment, I realized clearly that I needed to overcome the fear of combat insertions or I could suffer serious consequences of my own doing. It was an important lesson.

Removed from this incident by forty years, I'm more aware of the inner voices I heard that day than the actions we were involved in. The voices of authority and duty were telling me to get a grip, to be fully aware and to control the situation or wind up dead. It was Jarman's voice, the lifer from Ohio State ROTC days, telling me to "get with the program" or there would be unacceptable consequences.

The other voice seemed to be a mixture of my own fear and moral courage. It told me that this duty was irrational and unjust, and I had been coerced into something unnatural and illegitimate. The two conflicting messages presented a fork in the road that forced a choice: I would either cooperate to do what the immediate situation demanded, or I just might die.

I realize now that this incident was part of a mental process well described by Robert Mason in his book *Chicken Hawk.* In his story he identifies three distinct mental states that he experienced as a Huey pilot in Vietnam. His first 90 days (or so) in country was the first mental state, which was controlled by fear. His behavior during that period was overly cautious as self-preservation dictated his actions, leading to unreliability and limited effectiveness as a combat pilot. In my experience this corresponded to the months of May, June and July 1970.

Mason goes on to say that as one dangerous combat mission followed another, a belief of the futility of self-preservation developed, which led to the conclusion that death in combat was inevitable. Similarly, evidence of the futility was all around us as some pilots and crews died unexpectedly, while others did not. It seemed to be random. As a result, a second mental state that developed was characterized by cold, uncaring efficiency. The logic went something like this: "If it's likely I'm going to die here, then to hell with fear. I might as well perform my duty the best way I can so that others won't be victims of my fear-driven mistakes." In this second state, which Mason said ran for about six months, fear nearly disappeared and was replaced by a machine-like effectiveness. He became what the government sent him there to be: an efficient killer. For me this state ran roughly from August 1970 to R&R in late January 1971.

Mason's third mental state gradually evolved as he counted down the days to DEROS (Date of Estimated Return from Over Seas), or the go-home date. Just about everyone knew how many days he had left on his one-year tour of duty, and the exact date he would be sent back to the world. After 9 or 10 months of war duty, the mind began to speculate about the possibility of actually surviving, and this possibility produced hope which again led to self-preservation and fear. The third state was similar to the first, but in my case the fear in this state was not so much from the war as it was anxiety about returning home. (There's more about that in later chapters.)

Only a few weeks into my tour the opportunity for failure seemed exceptionally high, and at the rate I was going a lethal condition that could not be avoided seemed bound to develop. It seems likely that the mistake of turning the Huey's tail the wrong way was when Mason's stage two began to develop for me. Replacing fear with cold efficiency seemed to be the only safe choice.

That night 1LT Mitch Gregg, another pilot in the company, stopped by the hootch and dropped off a package from my wife containing candy, tape, a calendar and pest strips! I'd asked her to send the strips to deal with the bugs that flew in the door of the hootch at night when we left it open to get some fresh air. Al said that with the door open the place turned into "bug city." The pest strips were welcomed enthusiastically.

**Bad Shots**

Requirements in the motor pool ate up Saturday, 11 July and I didn't fly that day. Al and I went to the O Club for dinner and a movie that night; dinner was OK, but the movie was poor so we went back to the hootch to talk. A few minutes later Rick Mahan, Cat Brian, Mitch Gregg and Rob Denison came in to talk, and even though it was cramped in our small hootch, everyone enjoyed the conversation so much no one wanted to leave. We were becoming good friends.

By 10:30 our "guests" had left and Al and I resumed our discussion, which focused on math and physics. Being an engineer by training, he had more knowledge in these subjects than I did, so I listened to him talk about calculus, the relationship of energy and mass, pure energy in the form of light, the sun, atomic bomb technology, atomic structure, etc. I learned more in two hours listening to him than I'd learned in two quarters at OSU.

About 11:00 we heard an explosion, followed soon by two more, but they sounded so far away neither of us thought to break off our conversation until the siren announced a mortar attack was underway. We grabbed our helmets and ran for the bunker. The pilots who were there, including the CO, were entertained by my fashionable garb: helmet, white boxer shorts and flip flops.

It was the first time we'd been mortared in more than a month. There was no damage to our area, but one round landed on the far south side of the airfield and the rest landed between CTA and downtown Can Tho. The VC manning the mortars that night were bad shots, but we didn't complain.

**Holler & Ricks**

Martin Ricks and Dave Holler were warrant officers who arrived at the 191st together on Tuesday, 14 July, Bastille Day in France. They were nice guys, greener than grass as I had been when I arrived just two months before. The company was short of young warrant officer pilots, and these guys would replace some of the experienced ones who were due to leave in August. We were glad to see them.

Dave Holler was tall, maybe 6'3", and slim, with a square jaw, a mouth that usually hung open, and eyes with slightly drooping lids that made him look like a Bassett Hound. He was somewhat quiet and a serious pilot. Martin Ricks was friendly, about 5'9" tall with a moderate build and dark good looks. He too was quiet and a good pilot, but I never flew with him or Holler. Because they arrived at the same time and usually hung out together, they were often mentioned in the same breath as Ricks and Holler, or Holler and Ricks, as if they were a comedy team.

One day a bunch of us pilots were in a platoon meeting talking about various details of flight duty with Ron McConkey. I was sitting on the floor, leaning up against the wall next to Holler who was eating green olives from a can he had received from home. I had always disliked the taste of green olives, but sitting next to Holler the smell of his olives was enticing.

In our environment, almost everything that could be consumed ranged in quality from mediocre to unacceptable, which made the aroma of those olives mesmerizing. Although I didn't like green ones, I couldn't resist their aroma and asked Holler if he would give me one. He fished one out of the can with his fingers and handed it to me. Its center was stuffed with an orange-red pimento and together the colors reminded me of Christmas. I slowly placed it on my tongue and sucked the juice from it, then gently crunched it between my teeth, enjoying the flavor fully. It was unexpectedly exquisite, so I asked Holler for another one. He declined, unable to bring himself to spare a second one. After all, the can had been shipped all the way from the States, probably by his mother.

It's strange what minor details can be remembered from those days. But this story is all I remember of either Holler or Ricks.

**The Navy Way**

At 4:00 AM on Wednesday, 15 July, I was awakened to fly several ash & trash missions out of Eakin Compound in downtown Can Tho. Eakin was the MACV IV Corps Headquarters from which the operations in the delta were controlled, and all the brass (high ranking officers) that the company's second platoon, the Green Deltas, flew came out of Eakin.

By 2:30 PM we had been released from the mission and returned to CTA. I went to the motor pool to see what was going on there, but was unexpectedly called out to fly again, this time on a mission to Chau Doc to pick up four POWs and return them to Can Tho. Accompanying us to Moc Hoa was a Navy commander named Schellhase*, a cousin of Dave Schellhase* who was an All-American basketball player for Purdue University. When we arrived, we landed on a U.S. Navy barge that served as the barracks for the Navy unit that patrolled the river, part of what was referred to as the "Brown Water Navy," and stayed for supper.

From a letter dated Wednesday, 15 July 1970:

> *The Navy sure is alright. Gave us supper (had chocolate ice cream... really good-- and fresh baked bread...). They showed us all around the barge before we left. They sure know how to treat people!*

**Bounty Hunter Down**

By mid-July the company was back in the CA business in a big way. On Wednesday, 22 July, I was flying copilot again with CW2 Dave Armstrong in the trail position in our flight of five slicks following an early morning liftoff from CTA. We had made an insertion into an LZ and most of the crews in the flight were on edge having received enemy fire, but no one was hurt.

We climbed out of the LZ heading back to the PZ to get more troops. The air got cooler; the M-60s stopped firing. I was beginning to relax a little when Boomerang 6 called to say one of the Bounty Hunters had been shot down and had crashed on the edge of a canal near the LZ we just left. Armstrong immediately responded we would go, meaning we would rescue the downed crew. He asked Boomerang 6 for a compass heading to the crash site, and immediately tipped the nose of the Huey almost straight down, banking into the proper heading as we dived toward the ground. He leveled out just above the tree tops at 120 knots, max speed for a Huey, and told me to look out my window and tell him when we were directly over the downed Bounty Hunters.

The trees were zipping by under us as Boomerang 6 called to say the gunship crew was out of their Huey and receiving enemy fire. Armstrong asked what direction the fire was from, but 6 said he didn't know for sure. He asked what condition the crew was in, but 6 didn't know that either. Armstrong, his face grim and resolute, told me we would have to land between the Bounty Hunters and wherever the enemy fire was coming from to provide cover for the crew as they made their way to our Huey. Having no idea what was in store for us, I was wishing he had not volunteered us for this rescue mission that was unfolding faster than I could process what was going on. Armstrong said he wanted me to tell him exactly when we were over the downed Huey and what direction the nearest tree line was, which was the likely direction of the enemy fire. His speech was clear, direct, no nonsense. It fostered confidence that he knew what he was doing, which had a calming effect on me.

A few moments later we flew over our Bounty Hunter comrades and I told Armstrong we were directly overhead and the tree line was directly in front of us. The downed gunship had crashed on the edge of a small field on the narrow inside loop of a large canal. It was parallel to the water, straddling the canal bank with one landing skid nearly in the water and the other on the top of the bank, a couple of feet above. Its rotor blades were not turning, but there was no fire and the ship looked intact. I could see several of the crew members around the Huey down behind the bank returning the enemy fire, but that was all I could see as our Huey shot across the small opening, still at tree top level.

A couple of seconds later Armstrong said to the crew, “Grab something solid and hang on,” then pulling the stick all the way to the rear he put the Huey into a vertical climb. We were jammed back into our seats as the Huey’s nose pointed almost straight up, and nothing but blue sky could be seen through the windshield. I was instantly dizzy because of the vertigo caused from such a rapid and unexpected change of direction, and fought back the nausea that quickly followed.

The Huey climbed several hundred feet until the flailing rotor blades had no lift left in them, with the aircraft almost coming to a complete stop. This was as close to a hammerhead stall as I could imagine, and I instantly recalled my flight school instructor telling me that I would not learn to fly in combat until I reached Vietnam. At that time I had no visions of doing this, a maneuver that I thought the Huey was not designed for and nothing like anything taught in school.

As our climbing slowed to nothing, I began fearing that the Huey would start slipping backward toward the ground and all control would be lost. But at just the right moment Armstrong jammed the foot pedal to the floor and the tail of the Huey swung skyward as it did a complete 180 degree turn. Now we were looking almost straight down from several hundred feet up, the field next to the canal in clear view to our front, with the downed Bounty Hunters at the far edge of the field. We gained speed rapidly as we plunged toward the ground. With about a hundred feet left, Armstrong again pulled back hard on the stick and we started to level off, rushing toward the ground at unbelievable speed. My head was still spinning, and by now the acid taste of vomit hung at the back of my throat.

The whine of the engine grew louder as Armstrong pulled pitch trying to slow our downward rush. The Huey shuddered as its two big rotor blades desperately chewed the air, looking for enough lift to stop our rapid descent. We were again jammed into our seats from the force of gravity and I wondered if the rotor head could take this amount of abuse, or if it would just snap off from the rest of the Huey. My body was completely rigid, frozen with terror at the prospect of smashing into the ground at high speed, or not gaining enough lift in time to prevent us from mushing into the trees short of the clearing. I was sweating profusely, but Armstrong was dry as a bone, the picture of a man in control of the situation.

The Huey shuddered again ... and again ... as we neared the ground, but our descent was clearly slowing. Fifty yards from the field opening, Armstrong again pulled back on the stick and the Huey flared with its nose in the air, slowing our forward movement. By now the crew chief and door gunner were firing their M-60s into the enemy positions along with the second gunship.

Within seconds we were at a hover in the field next to the downed Huey and Armstrong told the crew chief to get the Bounty Hunters on board ASAP. The second Bounty Hunter gunship could be heard, mercilessly working over the tree line where the VC fire had come from with its rockets and machine guns, and the enemy fire had decreased substantially. The downed crew scrambled up the river bank and across the field into the back of our Huey. There were some bumps and scratches, but no one was seriously hurt.

I don't remember when I started to relax. It was sometime when we were back in the air en route to the PZ. What I had just been part of was impossible to fathom. I thought flight school had prepared me to be a good pilot: I had learned how to fly the chopper and to have confidence in my decision making. But there was no preparation at all for what Armstrong had done, and there was no way I could have finished what he had started if something had happened to him in the process. It was confusing and unsettling, and immensely humbling.

When we got back to the PZ, the rescued crew jumped out and we refueled. Minutes later we were back in the air with the other slicks, headed again for the LZ with another load of ARVN troops. This time the insertion was cold (there was no enemy fire), which is how it stayed through a number of additional insertions, and the extractions that followed later in the day.

Late that afternoon we reversed our early morning flight and headed back to Can Tho Airfield. And although I was totally wrung out, physically, emotionally and mentally, I knew I had been through a class with a master pilot and was grateful for a successful experience. No one in the company, except our crew, knew what we had been through that day, and no one cared. Each man in the Boomerang flight had his own story, which he would regard as his own special experience on 22 July 1970.

**The Perry Packin' Gals**

Friday, 24 July, I and Staff Sergeant Aaron Burley, the motor sergeant, flew as crew members to Soc Trang with Al and 1LT

Rob Denison, a pilot in the Green Deltas. SP4 Mike Tuke was the Crew Chief. We went there to see if we could round up parts from the 13th Battalion for the motor pool, and came away with a few. That afternoon we flew to Hotel-3 in Saigon to pick up a member of the 191st who was back from receiving psychiatric care in Hawaii. He was a notorious hard drug abuser, and apparently used LSD while he was gone which led to him going AWOL in Saigon. After wandering around the city in a stupor for a couple of days, during which he lost his bags, he turned himself in to the MPs. He was a mess; it was a shame.

The following night was the hail and farewell party for Major Sechrist involving champagne toasts and other formalities. It was too formal for Al and me, so we opted out by drinking too much Mateus wine. The evening was capped off successfully as we convinced Sechrist to let us take a Huey and crew the next day to explore an island off the west coast of the delta.

The best event of the day actually had occurred late in the afternoon when a package was received from the Perry Packin' Gals, a group of women volunteers from my hometown who sent goodies to Perry boys in Vietnam. The cache included six cans of cookies, peaches, fruit cocktail, potato sticks, Spaghetti O's, several bags of instant iced tea and Kool-Aid mix, popcorn, two pairs of socks, writing materials and a couple of paperback novels. It was a special treat from a group of ladies whose identities were never known, but I was grateful they were thinking of me.

**The Seabees of An Thoi (*ann toy*)**

One hundred twenty miles west of Can Tho, in the Gulf of Thailand, is an island named Phu Quoc (*foo kwah*) which was rumored to possess beautiful beaches of pure white sand that bordered a sea as warm as bath water. It was also said that the island was the site of a North Vietnamese POW camp which added some intrigue and danger to its image.

After wangling permission from Sechrist on Saturday night, Al and I and seven other guys headed for Phu Quoc the next morning at 8:30 in a Green Delta chopper. We had packed some life preservers, but no C rations or water and not knowing what to expect, we took off for parts unknown. The weather was beautiful as we flew to Ha Tien (*ha tee-en*), a city in Kien Giang province on the west coast of the Delta about 50 miles up the coast from Rach Gia. From Ha Tien it was another 30 miles to Phu Quoc, all of it over water, a new experience for us chopper pilots.

By mid-morning we had reached the bay at Ha Tien, and a few minutes later approached Phu Quoc. The sun, still behind us, was generating beautiful, white, puffy cumulus clouds around the island. As we got nearer we could begin to make out several deserted beaches, and the contrasting colors of light blue sky, white clouds, dark blue ocean and deep green foliage on Phu Quoc was spectacular scenery.

Not knowing where we were going, we headed directly for the middle of the island. Once we got to the coast we turned south looking for a promising beach, then flew almost to the southern tip of the island and found a place to land that looked like no one was around. Al was flying and he set the chopper down in a nice open spot not far from the water's edge that looked like it would be safe for the couple of hours we expected to be there.

When we stepped out on the grassy sand of the beach, the heat from the sun hit us like a hammer. There was almost no wind and it was sweltering, too hot even for bugs to be out. But it couldn't erase the beauty of the place. We ripped off our shirts, boots and socks and tiptoed down to the water's edge where the sea was quietly lapping the sand. I rolled up my pant legs and stepped into the water, which provided instant relief from the hot sand that was scalding the bottoms of my feet. The water was crystal clear and just as warm as expected.

Although the beach was strewn with cuttlebone, which I grabbed a sample of and stuck in my pocket, there were no fish, crabs, or any other sign of life. I walked out toward the open ocean until the water reached the bottoms of my rolled up pant legs, and turning round saw I had walked 40 yards or so into the water, which was not getting any deeper. The bottom was still pure white sand and the shallow water went out as far as I could see. I suppose a person could have walked out another hundred yards and the water would still not have been head high. It was gorgeous, but after a few minutes just too hot to stay out in. I walked back to the beach, grabbed my boots and socks and tip-toed back to the Huey's shady interior.

A short while later a young boy came over the sand dune behind us with a tin cup in his hand. We looked at him warily, but he had a big smile and looked harmless so we let him approach. He didn't speak English, but motioned us down the beach to a fresh water spring that flowed crisply out of the rocks and down to the sea. The cold water was a very welcome drink since we had neglected to take any water or food with us (duh!). After a few minutes we went back to quietly taking in the beauty of Phu

Quoc, mixed with short naps on the deck of the cargo compartment.

Not long after that, the dead calm was disturbed by the faint sound of an engine that grew louder as it came toward us. Looking across the lagoon we could see, to our great surprise, a motor boat pulling a skier across the calm water. We waved our arms and jumped up and down to get the boaters' attention, and the boat headed directly toward us at full speed until we could make out two passengers who were pulling the skier. It turned out they were Seabees from An Thoi, a village out of sight around the edge of the cove. They didn't seem surprised to see us, and wanted to know where we were from and why we were there.

We talked for a few minutes and noticed they had a case of beer in the boat, and asked if they could spare a few bottles. They said they couldn't, but would be willing to get us a couple of cases if we wanted. We eagerly agreed and forked over enough money for the beer, half expecting never to see them again. They roared off in the boat leaving us baking in the sun, but a short while later returned with two cases of warm Budweiser. With our thanks and good wishes they took off into the lagoon, towing the skier behind them. We watched as they went from one side of the lagoon to the other for the next half hour, towing the skier for awhile and changing skiers every few minutes.

The experience was similar to stories from *Tales of the South Pacific*, *McHale's Navy*, *M.A.S.H.*, or *Hogan's Heroes*. On a jaunt into the unknown we had unexpectedly run into Americans in the wartime Navy, and they were living la vida loca! And they certainly looked the part: bare-chested and bronzed from the sun, wearing shorts, flip flops and bush hats, roaring about the Gulf of Thailand in a speed boat. They had everything we wanted but couldn't get, and we envied them. How could they be living so well when we were regularly flying into live fire and losing pilots and crews in the process?

In a couple of hours the beer was gone and the sun was getting low, so we packed up and headed back to Can Tho. As we left we made a couple of steep, make-believe gun runs on the Seabees, passing over their boat by a couple of feet at 120 knots. They thought it was great fun and nearly fell out of their boat taking photos of us as we flew over. When we reached 1,500 feet, we radioed 191st operations to let them know we were OK and on our way back. It had been a good break from the daily grind.

That night we watched the movie *Midnight Cowboy* at the O Club, and the next day I mailed that piece of cuttlebone found on the Phuc Quoc beach to my wife. I had completely forgotten about it until I opened the envelope in which it was mailed, 42 years later!

## Bad to Worse

Revolving doors had been around since a patent was issued for one in the 1880s. It was designed to provide for a constant but limited flow of traffic into and out of a building while never fully connecting the inside environment with the outside environment. By the time of the American military effort in Vietnam, the revolving door had been adapted to warfare as an efficient method of providing a continuous flow of conscripted men into an illegal and immoral war against a people who had done nothing to deserve it.

Moreover, by limiting military service for a conscript to two years, and participation in Vietnam to 12 months, the public's objections to the war were kept under control while the American Government covered up the failing policies and tactics that would eventually result in total failure. The public at large slumbered through its duty and continued to send its sons into the breach, remaining as detached from the war as possible. And all the while the military services continued to train conscripts for set-piece warfare, not the guerrilla tactics used by the VC.

In my short time in the 191$^{st}$ I'd come to understand that the ultimate objective of every person there was not to win the war, but to get home alive. Both the clock and the calendar were on our side as the day for automatic DEROS would eventually arrive. But there was a huge cost to this practice, because each turn of the revolving door meant that men with experience and knowledge of how to fight and survive were being replaced with inexperienced people who didn't. This fundamental fact affected the Army's performance from top to bottom because everyone, regardless of rank, was subject to the revolving door that had become official U.S. policy.

The revolving door also applied to unit commanders, and on 28 July Major Sechrist was replaced by Major Blake Simon. There was a formal change of command ceremony which all available personnel not immediately committed to combat operations were required to attend. The battalion commander presided, passing the company's colors from Sechrist to Simon in proper military style, a tradition that was essentially meaningless to us hot, tired pilots. I was not fond of Sechrist and was glad to see him go, but I didn't realize what we were in for with this new commander.

Blake Simon would have been easy to miss in a crowd, even a small one. He was about 5'8" tall, but looked shorter because of his round-shouldered stoop. He had a brush cut which was covered by an Army baseball cap with a brim so flat it looked like it had been ironed. He spoke slowly in a plodding manner and had a distinctive lisp, an unfortunate characteristic for a commanding officer. His eyes, narrow set beneath a bulging forehead, had black irises that could not be distinguished from the pupils, reminding one of the eyes of a cartoon character like Little Lulu or Popeye.

The sum total of his personal characteristics led many to believe Major Simon was simpleminded, which led to the nickname "Simple Simon," used by the company's officers and enlisted men alike.

**A Strange Kind of Reality**

Life in the 191st was often like the stories on the TV series M.A.S.H. There were unusual events, interesting characters, ridiculous demands and unexpected situations. For example, one Sunday afternoon in July, I went to the central mess hall to have dinner and found a large, dead cockroach in the center of a serving of lime flavored Jello on my plate. It was disgusting. I confronted the mess sergeant about it. Instead of being apologetic, he was angry and said something about having nothing good to work with over there, and what did I expect, etc. He actually made me feel sorry for him and his plight, so I avoided eating in the mess hall again!

--- Ω ---

At the west end of runway 08 was a rickety wooden outhouse that the male Vietnamese employees on that end of the airfield used to relieve themselves. No one but Vietnamese ever used it. One day a Chinook, which had a tremendous rotor downwash that, when reflected off the ground, resulted in an opposite up-wash, slowly hovered over the outhouse. The up-wash lifted the flimsy outhouse into the air about 10 feet, and as the Chinook hovered away, it came crashing down in a mangled pile of boards. Within seconds the lumber pile began to move and a Vietnamese man, his black pajama pants down around his ankles, emerged from the wreckage, pulled his pants up and scurried away as if nothing out of the ordinary had happened.

--- Ω ---

Hygiene was always challenging because there was never enough time or resources to stay clean, and flying Hueys in combat was dirty, sweaty duty. To serve our daily needs, we had a small Quonset style building near our hootch that served as the latrine and shower room. It was perfectly adequate, except there often was no hot water.

Al loved to take long showers, but I couldn't figure out how he could tolerate cold water as long as he did. So one day I watched his routine. He left our hootch and went to the shower, then came out and fiddled with some equipment next to it. A few minutes later he came back to our room with a big smile on his face. I asked him what he was doing with the equipment next to the shower and he said he was firing up the boiler to get some hot water. Al had figured out that the boiler control unit would run for a few minutes and then shut down automatically, creating just enough hot water for one shower. I don't know why he didn't go into business selling hot showers to the other pilots. They probably would have paid for it.

**Enigmas**

In her song *Crazy*, Patsy Kline expressed a common human theme, and that is the futility, the craziness, of continuing to love someone when that love is not returned. It seems to be an unexplainable phenomenon, a product of complex factors so intertwined in the psyche that they cannot be untangled. She was singing about an enigma, and like Patsy we had our enigmas too, but they were not about love.

The Rules of Engagement I mentioned earlier could sometimes produce enigmatic situations when we were not allowed to engage those people who fit the profile of "enemy" in almost every way. The only missing criterion was weapons in the hands of those people. We knew they were bad guys and the inability to engage them produced confusion and anxiety.

--- Ω ---

There were other enigmas too. For example, on one occasion we were doing combat assaults into a hot LZ where the ARVN troops received intense small arms fire as soon as they got off the choppers. The rifle and machine gun fire went on throughout the morning and into midday, when we inserted more troops into the LZ. But an odd thing happened: right about noon many of the ARVN troops stopped fighting and began cooking their rice lunches over open fires in the LZ. It seemed crazy that anyone would pause to eat under such threatening conditions, but they did.

--- Ω ---

Another enigma involved monetary currency. When we arrived in country we were required to exchange our American dollars for MPC (Military Payment Certificates), an artificial currency created by the U.S. government as a substitute for dollars that we could use to purchase whatever we needed on any Army post or Air Force base in Vietnam. It was believed at the time that dollars should not be in the hands of the Vietnamese because U.S. dollars could be used as a universal currency to acquire weapons or other important materiel that could be used against us. Although MPC had purchasing power for the Vietnamese as well as us Americans, the Army would randomly require that we exchange existing MPC for replacement MPC. Consequently, all old MPC that had been acquired by the Vietnamese was rendered worthless.

The South Vietnamese had their own currency called the piastre (*pee-ass-tur*), which had relatively low purchasing power. They preferred having MPC which could be used to buy American goods. But having U.S. dollars was preferred above all because of their universal acceptance and high value.

American soldiers were carefully schooled about the currency situation and how American money in the hands of the Vietnamese could be used against us. Never-the-less, some GIs who did manage to get their hands on U.S. dollars would spend them liberally off post to acquire goods and services. The highly desirable dollar would buy more of anything than its MPC or piastre counterparts. It was very difficult to police and stop this behavior, and why Americans would knowingly engage in something that could do them harm seemed crazy. I suppose it's like smoking cigarettes. Everyone knows it can be deadly, but many do it anyway.

--- Ω ---

A Vietnamese expression for crazy was "dinky dau" (*dinky dow*), a slang term that most of us picked up in casual speech, especially when conversing with the Vietnamese women who worked as bartenders and waitresses at the Can Tho Officers Club. These young women, who sometimes dressed formally in silk Ao Dai, were attractive, and they liked to engage in light-hearted, flirtatious banter with the officers. (I'm sure they found this behavior produced the highest tips.) They spoke English well, but peppered it with Vietnamese slang, and the term dinky dau was often used. If an officer made a mistake paying for his food or drink, he was dinky dau. If he was inebriated and made a pass at a girl, he was dinky dau. Or, if he proposed in jest to marry one of the girls, they knew he was joking and said he was dinky dau.

What was truly dinky dau, however, was the war itself and the way in which it was conducted. Taking ground one day and then abandoning it, only to retake it later at the cost of more lives, was truly dinky dau. Moreover, the core strategy of trying to win the war by attrition (i.e. we kill more of the enemy than they kill of us) was also dinky dau, and literally reduced each GI and officer to the status of pawn, an expendable player in the deadly game of war. Why wouldn't this be considered crazy by any sane person?

The government did its best to portray this as reasonable, sound practice, continuing to hold up the specter of communism and the need to stop it as a valid reason for the insanity of such warfare. It was like Yossarian's situation in Joseph Heller's novel, *Catch-22*. You knew it was insane but there was no way out, and no way to reconcile what you were doing with Christian values. It was a cradle for PTSD (previously known as battle fatigue) which has destroyed many men's lives even though they survived their combat experiences.

--- Ω ---

As we soldiers were in 1970, Americans today are plagued by the problem of government dishonesty. What are we supposed to do when our government lies to us? If we just overlook it, it leads to more abuse of power and to greater dysfunction. But trying to stop it through the ballot box just doesn't work. So an intelligent person might conclude, as Patsy Kline said, that it's crazy, just dinky dau.

# Chapter 4 - AUGUST 1970

| *War* - Edwin Starr |
|---|

August 1st was my 90th day in country. Although a lot had happened in the previous three months, I still felt like a rookie, or a "cherry" as the GIs referred to new people, not battle-hardened. In spite of being ambushed and shot down, in spite of being decorated twice for valor, I knew I still had a lot to learn because every day and every mission involved unique problems that required their own solutions.

By now I was more or less acclimatized. The heat and humidity were still noticeable, but not as much so, and sleep was coming easier. The people I worked with and flew with were familiar, and a daily routine had developed. Most importantly, my understanding of the Huey and what it could do had improved dramatically and I began feeling confident about getting around the delta on my own.

A fairly good understanding of the company's operations had also developed: the Boomerang flight (i.e. the 1st Platoon) flew combat assaults or ash & trash missions, and the Rach Gia night mission every day; the Green Deltas (the 2nd Platoon) flew VIPs every day; the Bounty Hunters (3rd Platoon) either accompanied the Boomerang combat assaults, or flew the night mission.

But, I wasn't too sure what went on in the Wingnut Platoon (the 4th, or Service Platoon), and I had no knowledge about what went on in our sister company, the 162nd, or at battalion level and above, or how coordination occurred with Navy or Air Force units. At that stage, my world was limited to what went on in the Boomerang Platoon.

**Birthing a Lifer**

By this time, Burley, Pomona and the grease monkeys had made significant progress in the motor pool, and a special inspection of our equipment and operations was scheduled at Simon's request for Tuesday afternoon, 4 August. That morning, after a thorough cleaning of the motor facilities, equipment and grounds, I briefed Simon on the status of all the trucks. He seemed far less interested in that than he was in trying to intimidate me about the inspector's knowledge of motor maintenance. He told me I'd better "know my stuff" for the inspection, which was a joke considering what little I knew about truck maintenance I'd learned on the job.

Back at the motor pool we waited for hours for "Colonel Inspector" to arrive, hoping we could get his ear while he was there, and some help in getting more parts that were badly needed. But the Colonel never showed up.

I was not surprised, but Burley was mad and Pomona and the grease monkeys were pissed off. It had taken much time and effort to make the place presentable for an inspection, so it was not surprising that the anger and disappointment quickly turned to disgust, and then resentment. Regardless of whatever the excuse that was given for his no show, I figured it was not due to pressing war demands, but probably because he couldn't be torn away from a tryst with his mistress that day.

Personally I felt as though I was making progress as I could ignore these disappointments with the understanding they were just a game. I understood this was the lifer's game in which the first rule was: "Do as I tell you, kiss my ass, act intimidated by me and I will grant you a good efficiency report." For a lifer like Simon, just one bad efficiency report would spell t-h-e-e-n-d-o-f-y-o-u-r-c-a-r-e-e-r. Or, as the Vietnamese girls would say, fee-nee (for the French word finis, or finished)! The lifer's game was raw intimidation, equivalent to high school bullying. It was pure el toro poo poo.

Those of us who observed the game from a distance found much humor mixed in with disdain for the senior ranking officers who played it. Very base and descriptive phrases like, "do it or I'll jump in your shit," or "the major is coming, everyone jump through your asshole" were typical statements that could be heard from lower ranking officers and enlisted men. But the most derogatory expression of all came from one of the lieutenants in the company who would say on his way to the latrine, toilet paper in hand, that he was going to go give birth to another lifer.

**Arc Light**

We pilots had seen the huge craters that B-52s made when we flew over areas of previous bomb strikes, but I, personally, never had seen the explosion of 500 or 750 pound bombs dropped from many thousands of feet in the air. In early August that was about to change.

About 9 AM one morning we were southeast of Can Tho en route to a pickup zone. Earlier that day our Boomerang flight had combined with a flight of 162$^{nd}$ Vultures at CTA, resulting in a large flight of 10 slicks. The Hueys in the flight were communicating with the lead ship on FM radio, while lead was also on UHF and VHF radio to communicate outside the flight. I was copilot in a Huey in the middle of the formation.

None of us were comfortable being mixed in with the Vultures because it took us out of the tight five-ship routine we followed in CAs. In this situation, ten ships from two companies in one flight produced an added level of uncertainty about the combat assault we were about to execute, and no one was happy.

We had been in formation at 1,500 feet for about 20 minutes when the lead pilot called the flight on FM asking if anyone in the flight had heard someone say, "Bombs away." No one replied, so lead called a couple of the ACs in the flight and asked them if they had heard "bombs away," to which they replied they had not. A moment later all hell broke loose as the ground in front of the flight began erupting from the exploding bombs of a B-52 strike! We recognized immediately what was happening, as did the lead pilot, who had no time to give instructions for an orderly turn away from the strike. Like a flock of quail startled by a hunter, the flight instantly broke apart with Hueys going in every direction away from the danger, hoping to avoid running into a stream of bombs from this B-52 bomber strike, known as an Arc Light.

No one was hurt, but once we had made our way to the PZ and landed there were plenty of irate pilots wanting to know why C&C didn't know about the Arc Light that we nearly flew into. There was no answer then, nor any later, so by default the incident was characterized using an Army phrase that was used to explain the unexplainable: "10% never get the word." We figured we were in the 10% who didn't get the word about the Arc Light that day.

It was an uncomfortably close call. The image of a disciplined, orderly Huey flight breaking apart in panic right in front of me is still fresh in my mind's eye.

**Once a Dustoff**

There were helicopter units in Vietnam whose only duty was to evacuate wounded soldiers from the battlefield, usually referred to as medevacs (short for medical evacuations), and their radio call sign was "Dustoff." Medevacs were distinguished from combat choppers by the universal Red Cross sign on the sides and noses of their Hueys. I figured Dustoff pilots must have balls of brass to have the courage and stamina required to fly into hostile situations and pick up wounded or dead people day in and day out.

The 57th Medical Detachment, the first medevac unit sent to Vietnam in the early 1960s, was located in Soc Trang and was the nearest medevac unit to Can Tho. Under normal conditions wounded personnel picked up by the 57th's Dustoffs were evacuated to the 3rd Surg Hospital at Binh Thuy Air Force Base just north of Can Tho Airfield. The medevacs were bold and effective, but they couldn't be everywhere at once. Now and then a regular slick or gunship would have to provide medical evacuation if a Dustoff was not in the area.

One early afternoon we were shut down with the rest of the Boomerang flight at an airfield in the western delta. We were between morning insertions and afternoon extractions from a particular LZ we were working that day when a message came over the radio that a U.S. soldier nearby had been severely wounded and needed to be evacuated. Our crew volunteered to evacuate him, and when we arrived at the pickup he was loaded into the Huey in his poncho, bleeding profusely through the meager dressings someone had applied. Our crew chief that day, Sergeant First Class (SFC) Jones, the company supply sergeant, was flying with us to meet his flight requirements to stay on active flight status. He immediately began dressing and redressing the wounds.

Once off the ground we headed for 3rd Surg as fast as we could go. But even at that speed we knew it would take 20 minutes to get there. With the doors open and wind whipping through the chopper, there was blood flying everywhere, and soon the inside of the windshield was dotted with dried, coagulated red splotches. The AC was flying, so a couple of minutes into the flight I turned and looked back at the cargo floor. It was a scene of carnage as the floor was covered almost completely with blood, as were SFC Jones and the door gunner. The poncho on which the soldier was lying was flapping wildly in the wind and I could see his face each time the corner of the poncho lifted. His eyes were open, but his face was ashen and he didn't look conscious.

Jones told us he could detect no vital signs, but we kept the Huey at red line anyway hoping we would get to 3rd Surg in time to save him. We felt desperate, frustrated at the slowness of our flight. I wanted to get out and push, but all I could do was hope. So I hoped, and hoped some more.

Finally the AC told me to look up the radio frequency for 3rd Surg and call them to be at the helipad to expedite a pickup when we touched down. A few minutes later we spotted the red cross on the hospital helipad and made a rapid approach and landing. Hospital personnel were waiting, and as soon as our skids hit the ground they whisked the wounded and his poncho onto a jeep and headed for the hospital emergency room.

The cargo area was a mess. There was blood and small pieces of flesh everywhere and Sergeant Jones was covered with blood from his shoulders to his knees. The door gunner was nearly the same. We made our way to Can Tho Airfield to refuel and clean up before returning to join up with the rest of the flight. Jones and the gunner changed clothes as the AC and I washed out the chopper. We had plenty of water to flush away the carnage; but nothing could flush away the images. Not even 40 years does that.

The path we crossed with that GI that day, a man whose name we never knew, was gruesome. It was especially disappointing when we learned that, in spite of our efforts, he didn't survive. But we had tried to preserve life, not destroy it, and for that opportunity I was grateful.

My deepest respect has always been reserved for the everyday Dustoff pilots of Vietnam. They may have been men with balls of brass, but they had hearts of gold.

**PIC**

My first mission as PIC (pilot in charge) was the night MAJ Simon flew as copilot on the Nighthawk lightship, probably during the week of 3 August. Our flight had been given a free-fire area south of Rach Gia that, like many previous nights, was very active with VC. Flying C&C, I coordinated the operation at 1,500 feet where we were out of the range of ground fire, but the gunship and lightship got into a hornet's nest and took 20 hits (rounds) of VC fire between them. The Bounty Hunter copilot was wounded in the thigh, and we flew him to 3rd Surg where he was patched up and later declared to be OK.

## Bronc Buster

I have forgotten many of the enlisted men I flew with, but a few stand out from the crowd. One of them was a muscular fellow named Overstreet, who was about 5'6" tall, although he seemed much bigger than that. Overstreet was from Wyoming and in his civilian life he'd been a rodeo rider, a real live bronc buster. He seemed to have muscles on top of muscles long before that look was made popular by Arnold Schwarzenegger. He was the only crew chief in the company I knew who used snuff, and the packed fullness of his bottom lip stuck out when he had a pinch in.

Overstreet was smart and strong, and could man-handle an M-60 machine gun with one arm. I recall looking back at him while we were on approach to LZs and seeing him with a smoke grenade in his teeth and a hand grenade in one hand, while he fired the M-60 with his other hand. I asked him why he did that, and he replied that you could never know when you were going to need a smoke grenade to mark a target, or a hand grenade to toss at VC in an LZ. Armed that way, he was always prepared for whatever might develop. He may have been short of stature, but he was a man's man, and early on I decided that if I was to be shot down again I would want Overstreet with me. He was tough as nails.

## Monsoon Surprise

Monsoon season was fully upon us and the delta's rice paddies were nearly full as Gordon Clayton and I flew the Nighthawk

C&C again on 7 August. During the afternoon and early evening we worked over a free-fire area south of Rach Gia with the Boomerang light ship and a Bounty Hunter gunship. But earlier than usual it turned pitch dark, owing to a solid cloud cover. No moon shown; the air was murky and heavy with humidity; visibility was very poor, but we could see ground lights here and there from the air.

We refueled at Rach Soi airstrip and returned to the free-fire area to continue coordinating fire support for several outposts that needed help. Eventually both the lightship and gunship left to refuel and we were alone to support the ground troops. However, there was little we could do with our meager firepower.

As we circled the area focusing on what was happening on the ground, we paid little attention to the weather and how it was deteriorating. As Clayton was at the controls, he made a sharp left turn toward the coastline, directly into a heavy rainstorm that had crept up upon us with no thunder or lightening as a warning. At once we were plunged into complete darkness as violent wind gusts started to hammer the Huey. Peering out the windshield to the west we recognized lightening flashes, and then realized we were in a thunderstorm.

We had been warned in flight school to stay away from thunderstorms because Hueys were not built to take the punishment a storm could dish out. But here we were, unintentionally violating the rules, getting knocked around like a toy at 1,500 feet, glued to the instruments and wondering if and how we were going to get to safety.

Clayton wrestled with the controls and I searched for lights on the ground, or anything else that would give us assistance to get away from the storm. As we turned eastward, the direction from which we had come into the storm, I peered through the gloom beneath us and found a narrow hole in the clouds that went all the way to the ground, more than a thousand feet below, where I could make out a few lights. I told Clayton about my find and he quickly decided to make a tight spiral down the hole until we came out of the storm. Banking sharply to the left he began the descent, getting kicked all over the place in the process by the winds in the clouds.

I kept wondering what was holding the rotor head on the mast, figuring it could be ripped off at any moment from the wind gusts that were hitting us. We dropped a hundred feet, then 200, then 300, and the Huey was still getting belted. The rain was intense and relentless as it whipped through the open doors of the cargo area and drenched the crew chief and door gunner, who could not close the cargo doors in time to avoid a bath.

The altimeter said we had lost 400 feet, then 500, and still there was no let up. I glanced at Clayton who was glaring at the instruments and struggling with the controls as he fought to maintain the tight spiral descent. It was a maneuver made up on the spot that resembled nothing like what we had been taught, but it was working. The lower we got, the more ground lights I could see until finally we broke out of the clouds at around 300 feet above the flat delta terrain, the rain still pounding the windshield and the wind gusty, but lessening.

We could now see the ground and where we were going, so I grabbed the map to see if I could figure out where we were, but there was nothing recognizable under my red map light. We flew toward a dim cluster of lights a half mile away and eventually the rain subsided enough that we could see a small landing strip below us. We lost altitude quickly and made a landing at the airstrip where we found a JP-4 bladder and filled the Huey with fuel. Within minutes the worst of the storm had passed and we took off for Rach Gia to rejoin our comrades who had weathered the storm on the ground.

Nothing much was said about our harrowing experience with the storm and just how close we had come to disaster. I guess we felt the less said the better, because this was all in a day's work. Plus we didn't want anyone else to know we were dumb enough to fly into a thunderstorm! Back on station we went ahead with our support of the outposts, and it turned out to be a long night. I logged 15.5 hours of flight time, more hours for me than any other single day in my Vietnam tour. It was memorable, but chilling, and I wondered just how many close encounters with disaster one could have without paying the ultimate price.

**Bloody Mary (of Moc Hoa)**

We had been in or near Moc Hoa, situated a few miles west of the Parrot's Beak near the border of Cambodia, many times. Located in Long An Province just north of the Plain of Reeds, Moc Hoa was very rural and the peasants very poor. It sat on the edge of the Vam Co Tay River, straddling a Viet Cong infiltration route from Cambodia. Consequently, it had strategic importance for both the VC and the Americans.

During the summer of 1970 there was a concerted U.S. and ARVN effort to restrict the infiltration route and thus strangle the supply of arms and ammunition making its way into South Vietnam from Cambodia. The 191$^{st}$ was tapped for troop support for these operations and we spent many days in and around the

Moc Hoa area doing combat assaults. There was occasionally a fire-fight in a hot LZ, but most of the time the LZs were cold and the flights became routine.

Around noon each day we would take a lunch break and eat our C rations, and try to ward off the hoard of 10-15 kids (almost always boys) who would inevitably find us and clamber for chocolate, or chewing gum, or cigarettes. With these kids we had to be most careful, not so much for their safety as for ours, because stories abounded about Vietnamese kids carrying live grenades to blow up Americans, and we could never tell if the Moc Hoa kids might be carrying something lethal.

Generally speaking the kids were fine, mostly just curious and happy when they got some treat from the Huey crews. But caution was always required, because even if they weren't deadly they would steal anything they could touch, like pens or pencils, watches, C rations, sunglasses or anything else that wasn't nailed down. (Moc Hoa was where the Red Box had disappeared from our Huey.) I was not sure how to handle this situation, but I took my cues watching the experienced pilots and how they handled it, and I eventually relaxed with these young ones who made up a good portion of the "local color" we observed.

But there were other aspects of local color much different from the kids, and one of them was a haggard old woman nicknamed "Moc Hoa Mary." Mary was probably in her fifties and nearly toothless, with a red stain down her chin from chewing betel nut, a mild pain killer. She was frequently around the choppers when we were parked and waiting for our next mission, and most of us knew she was trying to sell herself, probably to make a living. Most of us stayed away, but not everyone.

One day while we were shut down one of the warrant officer pilots decided to take Mary up on her perpetual offer and had relations with her in the back of his Huey. A few days later he was under treatment for a sexually transmitted disease (STD, or venereal disease) that Mary had been pleased to pass along with the fun she provided. To the rest of us this pilot had sunk pretty low by sampling the local color of Moc Hoa too closely.

STD was not the only scourge to be avoided, as there were other more dangerous and less curable afflictions awaiting lonely or unsuspecting Americans. One of them was called amoebic dysentery, a form of severe diarrhea usually caused by fecal contamination, and one of the 191st pilots contracted it.

For several months this pilot, whose name I can't recall, was in miserable shape, unable to shake the problem and yet not sick enough to be taken off flight status. He could barely control his bowels and we sometimes saw him bail out of his Huey in an LZ and run out behind the chopper to quickly relieve himself, then jump back in and fly off with the rest of the flight. He lost a lot of weight, maybe 40 pounds, and eventually transferred to the 2nd Platoon where the flying was less demanding, and he had better access to toilets.

**Farewell and Adieu**

Saturday night, 8 August, was very difficult, not because of a tough mission but because we were saying goodbye to eight "old guys." Leaving was the Executive Officer (Ted Everest), one Bounty Hunter pilot, two Green Delta pilots, and four

Boomerang slick pilots, plus a serious amount of combat experience. My hero, Dave Armstrong, was one of the departing Boomerangs and the thought of not having him around was depressing. It was a sad evening, even though we all were happy for them that they were going home. They had earned it.

The guys who were leaving each got a painted and engraved Boomerang trophy (a real Australian boomerang) and each said a few words. Then it was time to belatedly, but formally, welcome the new CO, MAJ Simon. He was first doused with beer (a ceremony called "meeting john") and then carried, against his will, to the post cesspool where he was thrown in. Boy, did he stink!

**Fantasy Land**

It wasn't as if there were no women around, but the women we saw everyday were Vietnamese. They were often attractive (even very attractive), but not "dames" as we defined the word. There were shows at the officers club with lusciously endowed Philippine dancers, also very attractive, but they were not dames either. Based on our own bias, we thought, as did the sailors in the musical, *South Pacific*, that dames had round eyes, not almond-shaped eyes, and that they spoke with New England, or Southern, or Midwest accents; not the broken English of the Vietnamese.

We all wanted to go home, of course, and leave behind this hellish war. But what we longed for even more was the tender touch and the loving affection of a dame, a round-eyed American woman. My American woman was with me in the letters, photos, cassette tapes and packages she sent, but she was not there to hold and caress me, to make me forget the dark uncertainty of death and the despair that existed all around.

Now and then, a moment of relief could be found in this harsh environment. Sometimes it would occur in a group with the other guys and our guy activities. But sometimes that moment came at night, flying at 1,500 feet, listening to Karen Carpenter's smooth voice singing "We've Only Just Begun" on AFVN radio.

| *We've Only Just Begun* - The Carpenters |
|---|

And at other times, if we were in the area we could hear the female tower operator at Sanford Army Airfield. The first time I heard her voice was on 13 August on a flight with Connie Leopard to get motor parts at Long Binh. A couple of miles south of the airfield we dialed in the Sanford Tower radio frequency and requested landing instructions:

"Sanford Tower, this is Boomerang 26 about two miles to your south requesting landing instructions, over."

What came back was astonishing! A soft, sexy, luscious voice in perfect American English whispered:

"Boomerang 26, this is Sanford Tower. The wind is from the west at five miles per hour and the barometer is two-niner-point-niner-six. Please announce your entry into the traffic pattern for a landing on runway two-seven, over."

We looked at each other. Instant fantasy!

We didn't know anything about her, which fueled our imaginations, and yet we wanted to be near her, a round-eyed woman! A dame! Here, in war-torn Vietnam! And we could actually talk to her! In fact, lots of pilots talked to her every day, luxuriating in the warmth of her voice. But they were always careful. No one dared say anything vulgar or disrespectful because no one wanted the voice and what might be behind it to go away.

There were, of course, many questions associated with our fantasy:

What did she look like? Was she tall or short? Blond, brunette, or red-head? Young, or old? Plump, or slim? The mind raced!

And what was she really like in person? Was she passive, or aggressive? Demure, or brazen? Intelligent, or not? Outspoken, or reticent? The anticipation was enchanting!

And who, by the way, had the nerve to allow her to talk on the control tower radio with that voice to a bunch of companion-starved chopper pilots?

The questions were almost as endless as the imagination!

Pilots from all over southern Vietnam had either heard the Sanford Tower operator, or heard of her. And, of course, no one who talked to her wanted his fantasy bubble to burst. She was a celebrity and we all supposed she could be the girly, womanly, female, feminine dame that could cure whatever was wrong with us by just being near her.

But then, one day the inevitable happened. A foolish pilot landed his Huey, crossed the airstrip and climbed the tower stairs to see if he could get a glimpse of our "fantasia." One look and his dream was crushed, then to the world of aviation he disclosed that she was a short, plump, dishwater blond in rumpled fatigues and not the babe we had in our minds' eyes who would cure our blues. We all knew the truth was going to come out at some point, but when it actually did American hearts were broken all over Vietnam.

--- Ω ---

Long Binh, where Sanford Field was located, was itself like a fantasyland, a sprawling complex of office buildings, recreational facilities, barracks, warehouses and concrete. It was home for the headquarters of USARV (United States Army Republic of Vietnam) and completely secure, sometimes referred to as "the nest." It's where General William Westmoreland and his successor, Creighton Abrahms, ran the war and the place where one of my high school classmates worked in military intelligence.

Sanford Airfield was also the home of the 120th Assault Helicopter Company that used the radio call signs Razorbacks and Deans, and while we were there looking for parts I checked to see if any flight school classmates were in the company. Sure enough, Don Basil's name was on their pilots' list so we waited around a few minutes for him to return. In the meantime I saw CPT Gene Hoffmeyer there, another pilot from my flight school class, and we talked very briefly (Gene was a pleasant guy but somewhat taciturn, so the conversation was short). When Don came back we talked for a few minutes and I learned that his father worked for the PX system in Saigon, and they got to see each other every week. Lucky guy! All of my family was 12,000 miles away, but his father was almost next door. What a strange war it seemed to be!

Leopard and I went to the parts supply office and found that we needed all kinds of paperwork we didn't have, so we took the papers with us back to CTA. We returned the next day with the completed paperwork and collected some badly needed Huey parts, but were disappointed because the tower operator that day was a guy!

### The Parking Lot

Each helicopter company at CTA had designated parking places for their aircraft. Huey companies had concrete revetments on the flight line near their maintenance hangars. When pilots landed, they would request clearance from the tower to hover to their parking places using a nickname for the location that corresponded with their company call sign.

For example, the 191st Boomerangs would ask to go to the "racks"; the 162nd Vultures to the "roost"; and the Cavalry Dark Horses to the "stables"; all well known locations on the airfield. While hovering at the airfield I heard a few, more interesting, nicknames: the 271st Innkeepers asked for clearance to go to the "pub," while another Chinook company called the Bartenders asked to go to the "pisser." And, of course, the 235th Satans requested clearance to go to "hell"!

### Sol

Early the week of August 17th, Ron McConkey was relieved as the commander of the 1st Platoon, so as assistant platoon leader I automatically took over in his absence. When I asked MAJ Simon why McConkey was gone, all he would say was, "You're the platoon leader now and you'd better get things straightened out." Just exactly what things he had in mind to straighten out were unknown to me, so the situation seemed similar to my assignment as motor officer when Sechrist told me to clean it up, with no specific instructions.

I was uncertain about being in charge of the platoon, but not afraid of the new responsibility. I'd made a lot of friends among the platoon's pilots in the last 90 days, and knew there were a lot of good, competent people in the slick platoon who were eager to have good leadership. There was a lot of cooperation and I regarded the situation as a positive challenge. All I had to figure out was what McConkey had done wrong so I wouldn't repeat it.

Four days later, on Friday, 21 August, my short-lived assignment ended when a captain named Henry Solomon arrived in the company and became the permanent 1st Platoon Leader. He was from Miami, Florida and on a second tour of Vietnam, arriving in Can Tho from an 18-month assignment in Panama. On his first tour he had served as a pilot in the 101st Airmobile Division in I Corps, and we all soon learned that he was nobody's fool.

Sol, as we called him, was the only black pilot in the company, but that difference in skin color was meaningless. All we cared about was his competence and his good character, and he had both. He was quiet, respectful, friendly and positive in his outlook, all qualities that were attractive to the pilots and crews in the platoon. But he also had a casual aloofness that made me uncomfortable, and I didn't have enough time in the platoon after he arrived to resolve my discomfort. The bottom line was I liked and respected him, and believed the revolving door had turned in our favor when Sol walked through it.

--- Ω ---

The old guys were leaving and needed to get rid of what they had accumulated, so on Saturday morning Al and I bought their black and white TV and antenna, a refrigerator and a record turntable. Later that day a group of us put the TV antenna up atop a 30' pole next to our hootch in a perilous exercise that resembled a slapstick comedy routine. Somehow no one was hurt.

In spite of reservations about how good the TV reception would be, it turned out to be quite impressive. However, most of the programs we got were largely Vietnamese TV stations. But there were American TV programs too, and even though there were so many other distractions (like sleeping, the O Club, night flying duty, etc.) we managed to see reruns of *Bonanza* and *Gunsmoke* now and then.

**Company Social**

Once every three or four months there was a company-wide party for all the members of the 191st, and the menu was usually steak, crabs and ice cold beer. I had no idea where the steaks came from, but I figured they might have been traded for helicopter or truck parts. The salt water crabs were fresh and picked up by some of our pilots who bought them out on the west coast near Rach Gia. The steak was cooked over an open grill and the crab boiled in a large pot, and the combination of aromas really made the mouth water. It was a welcome change from cold C rations in the field, or ham sandwiches and hard boiled eggs at the O Club.

These company-wide parties were always held outdoors in the late afternoon next to the basketball court that was close to the company headquarters. The opportunity to rub elbows in a relaxed setting with people we didn't see much helped to moderate dissension that arose between individuals, such as that between me and Blake Simon regarding McConkey's departure. And they also helped close gaps between groups and factions in the company that resulted from differences in rank, status, culture, intelligence, or color. For example, there was a group of four black men who formed a clique called "The Brothers" who hung out together and separated themselves from their peers. Why they did that I didn't know, but all of them had ground assignments and were not on flight status. Maybe that was the common factor among them.

SP4 Tubbs, who worked in the orderly room, was one of the Brothers. He was quiet and friendly, and while he chose to segregate himself he was neither confrontational nor disrespectful. That afternoon at the party he was hanging out with the rest of his friends waiting for the steak and crab to be served. Out of boredom, I guess, he joined me on the basketball court to pass the time shooting baskets.

Tubbs had all the physical abilities a basketball player could want. He was fast to the basket, handled the ball well and knew how to keep it away from opponents. I was no match for his skills, but just for something to do we started a one-on-one game of hoops. It was blistering hot with the usual high humidity, but we ignored the heat and went at each other for about 30 minutes until we were dripping and our clothes were saturated with sweat.

When Tubbs got the ball, he drove past me for a layup that I couldn't block. When I got the ball I'd shoot it from 25 feet away and sink it most of the time. Each time I did, Tubbs would just shake his head like it was a lucky shot, until we got deep into the game and the score was close. Then we both realized we could do this all afternoon. He couldn't stop me, and I couldn't stop him. It was essentially a draw that we couldn't break.

When we quit, we shook hands and had a beer and some steak. He went back to his clique with the Brothers and I sat with Al and some other pilots. But the goal for this company picnic had been achieved in one small way: some respect had been built between an officer and an enlisted man, between a black man and a white man. And the by-product was trust.

--- Ω ---

The Boomerang flight was working about 30 miles southwest of CTA on Monday the 24th and landed to refuel at Vi Thanh (*vee tahn*) airfield which had a lot of refueling points. The unit refueling behind us was a cavalry unit called the Comanches, which I knew was the unit Jack Wyatt was in. (Jack was a flight school classmate.) I walked back to the lead aircraft to ask if Jack was in the flight, and found him sitting in the front seat of the lead Huey. When he saw me he jumped out, a broad grin on his Texas face beneath crinkled eyes and a hooked nose. We shook hands and exchanged what little news we had of people back home and people we knew were in Vietnam. Even after three months in Vietnam, it still amazed me that Wyatt, Yost and I, who had shared the same house in Savannah, were stationed just 30 minutes from each other by air.

--- Ω ---

Wednesday night, 26 August, eight pilots attended a briefing on a top secret mission that would occur over the next two days.

Walt Blocker, Matt Akin, Connie Leopard, Mitch Gregg and I attended from the 191st and found out we would be flying to Phnom Penh, Cambodia to pick up the U.S. vice president for a special diplomatic mission. It was logical that high ranking officers and their crews would fly this mission, and that our Green Delta pilots, who were accustomed to flying VIPs, would be involved as well. It was not clear why Blocker and I were selected because he was a gun pilot and I a slick pilot. Our experience was in combat operations, not diplomatic missions, but I was flattered to be involved and looked forward to what the next two days had to offer.

**The Phnom Penh Mission, Background**

By the middle of the nineteenth century, France had colonized portions of Southeast Asia and formed a federation of colonies known as French Indochina (Indochina derives from the French word *Indochine*, a combination of the names India and China). Cambodia became part of Indochina when its king requested help from France to protect it from Thailand to its west and Vietnam to its east, which had claims on Cambodian territory. In 1863 Indochina became a French Protectorate and remained under French control until 1954 when France lost the Indochina War and withdrew its military from Southeast Asia.

Phnom Penh (*nom pen*), a city of more than 400,000 people in 1970, was founded in 1434 at the confluence of the Tonle Sap, Mekong and Bassac Rivers, and was designated the formal capitol of Cambodia when it became part of Indochina. As the capitol of the country, Phnom Penh became the center of economics, industry, politics, culture and diplomacy, and the combination of Cambodian and French architectural influences led to its nickname "Pearl of Asia." The Cambodian people, known throughout Southeast Asia as the Khmer (*kuh-mare*) were equally as beautiful as their city, and known to be gentle, generous, humble and hospitable.

In the 1960s, Cambodia was formally neutral and had refused to choose sides in the American War in Vietnam. As the war escalated, both the VC and NVA began to use Cambodia as a safe haven from which they could attack U.S. and South Vietnamese troops and installations, withdrawing to Cambodia whenever safe sanctuary was needed. The famous Ho Chi Minh Trail that was used to provide war supplies to the VC and NVA from North Vietnam had extensions that ran south into eastern Cambodia, and by the end of the 1960s the need to eliminate the enemy hiding places and supply routes had become a top priority.

In March 1970 the king of Cambodia, Norodom Sihanouk, was deposed by the National Assembly and the prime minister, Lon Nol assumed control of the new Khmer Republic as president. That development resulted in the communist rebel forces, known as the Khmer Rouge (*kuh-mare roozh*) openly assaulting the U.S.-backed Cambodian government. On May 1st, just about a week before I arrived in Vietnam, U.S. and South Vietnamese ground and air forces crossed the border into Cambodia in the Parrot's Beak area. The Cambodian Incursion, as it was called, was viewed as a major and overdue strategic development.

In spite of the incursion, by August the Khmer Rouge had attained control of a large portion of the country and had surrounded the capitol on three sides, threatening to cut Phnom Penh off from the outside world. Cambodia was quickly falling to the communists, so U.S. Vice President Spiro Agnew was sent to Phnom Penh as last ditch diplomacy to show support for what was then a sinking regime.

**The Phnom Penh Mission, Day 1**

On Thursday morning, 27 August, we began preparing for the two and a half hour flight we would make directly up the Bassac River to Phnom Penh. The overall plan was to fly to Phnom Penh on the 27th, and then on the 28th meet Air Force 2 and fly Agnew to the Royal Palace for a meeting with Cambodian President Lon Nol. We expected to be back at CTA the evening of the 28th.

An individual helicopter load is referred to as chalk. Our four Huey chalks were arranged in this order: chalk 1 (lead Huey) pilots were COL Kenzel and another pilot from the 307th Battalion; chalk 2's crew was COL Delips (don't know his unit) and CWO Matt Akin (one of our Green Delta pilots); chalk 3 was CWO Wayne Crocker (one of our Bounty Hunters) and myself; and chalk 4 was CPT Connie Leopard (Green Delta platoon leader) and 1LT Mitch Gregg (Green Delta pilot). Each Huey also had its crew chief and door gunner on board, making a total of 16 people in the flight of four Hueys.

Connie Leopard had been recently promoted to Captain and was an Aircraft Commander. He had been the supply officer when I arrived at the 191st, but had stepped down from that job when he became the Green Delta platoon leader. Leopard was about 5'10" tall with thinning hair and a slight southern accent that may have originated in Texas (although I don't remember where he was from). He looked to be about 35 and had been through OCS before flight school.

Invariably friendly and intrinsically good natured, Leopard loved to tell funny stories while sipping good whiskey. I liked him because of his simple and straight-forward manner, but he was just a bit too easy-going for my comfort, and didn't show the seriousness I expected in a combat pilot. He and Bob McConkey had got on well and seemed to enjoy each other's company on their frequent trips together into Can Tho City.

Leopard's copilot was a burly lieutenant named Mitch Gregg. About 5'10" tall with a black brush cut, Mitch looked like a football lineman. He was happy-go-lucky by nature, and when he consumed a little too much beer a goofy side of his personality came out. I loved Gregg, as did just about everyone else. He was a person you wanted in your corner when things got rough. I was glad he was going up river with us.

Walt Blocker was tall, lanky, quiet and known to be a solid pilot. We did not fly together and didn't hang out together either, so Walt was as much a mystery to me as I probably was to him. Matt Akin was an experienced Green Delta pilot who I never flew with and didn't know at all. I have completely forgotten what he looked like.

Our detailed plan for the trip was to fly directly to the Phnom Penh International Airport which was five miles west of downtown Phnom Penh, refueling once along the way at Chau Doc. Our preflight briefing revealed that the airport was closed to all air traffic because the Khmer Rouge were so close to the city that it could be overrun at any time. Since no one was certain just what the conditions would be like when we got there, the plan also included taking 55-gallon drums of JP-4 jet fuel and a hand operated fuel pump in case there was no jet fuel available when we arrived. We were also instructed to pack plenty of machine gun ammunition and C rations in preparation for whatever conditions we might find.

The morning of the 27th we went to the flight line to preflight and load up with supplies. There were 4 drums of JP-4 already in the cargo compartment and strapped to the floor of our Huey, along with several extra boxes of M-60 ammo and a couple of cartons of C rations. The whole helicopter smelled like JP-4 from some of the fuel that had spilled on the floor during loading. But if we and our Huey didn't smell pretty, we would at least look pretty because we had fresh flight uniforms to change into, and soap and shaving gear with us too. (The higher-ups didn't want the vice president to get the idea that the war prevented us from being presentable.)

It was early afternoon by the time we cranked up our nearly brand new Green Delta Huey, joined the three other ships in the flight and were in the air. We reached Chau Doc 45 minutes later and landed to refuel. In 15 minutes we were back in the air headed for God knew what. We crossed the border into Cambodia following the Bassac River knowing it would lead us directly to Phnom Penh, the only challenge being to find the airport once we got there.

Within an hour monsoon rain had caught up with us, but we had no trouble finding the airport. Soon we were at a hover moving toward the terminal building. It was 6:30 PM, yet the entire airport was deserted; a few parked planes, but no vehicles and no people were in sight. As we neared the terminal a Cambodian soldier came out waving his arms, and in the still-pouring rain told us to hover around the side of the terminal where we would find a place to unload our drums of JP-4. He said someone just beyond the chain link fence would be there to help unload the drums, which weighed more than 400 pounds each.

Following chalks 1 and 2, Blocker hovered around the building where we found a long, six foot tall fence that ran from the wall of the terminal out to the edge of the airport. The first two chalks successfully hovered over it in spite of the rain having increased substantially and the wind whipping around the terminal building in gales. Then it was our turn.

Blocker turned the Huey parallel to the fence, which was on our right side, and pulled pitch to get over it. At 10 feet in the air he started sliding the Huey to the right over the fence, just as a huge gust of wind hit the tail boom from the left side, pushing it quickly to the right. As the tail boom swung right the nose abruptly moved left and Blocker pushed hard on the right pedal to straighten us out.

With heavy rain pelting us and wind gusts mixing with the Huey's rotor downwash, the helicopter became a wet bucking bronco, jumping unpredictably this way and that. The wind gusts and the extra weight of the JP-4 drums in the back were straining the Huey's engine to the maximum. Blocker had a look of determined concentration as he made rapid adjustments to the controls trying to keep the Huey from settling onto the fence, which we now straddled.

When another strong gust hit the tail boom, Blocker pushed the entire left pedal until there was no more to apply. The nose moved to the left a bit, but then just stopped, and we found ourselves straddling the fence at a 45 degree angle. With all pedal gone and the engine power maxed out, a big burst of wind hit us from the top and we began to sink toward the fence below us. Blocker now wore an expression of deep concern as we sank a foot, then two, with the belly of the Huey now only a couple of feet above the top of the fence. The normal high pitched whine of the engine had decreased a bit, and I could hear it straining to keep up with the excessive demands of the situation.

Having an engine failure flashed through my thought, followed by the image of settling onto the fence and tipping over. In my mind's eye I could see the main rotor blades hitting the asphalt pavement and splintering, throwing jagged pieces in all directions. I could see the Huey's nose punching the ground as the two plexiglass chin bubbles and windshield shattered. I saw the four fuel drums break loose crushing me and Blocker in our seats, then rupturing and spilling fuel onto the pavement. I saw a huge ball of fire consuming the Huey and its contents as the engine ignited the spilled fuel. No one survived. It was an imagination out of control.

The sound of Blocker's voice brought me back into the moment. The Huey continued to slowly settle toward the fence, its power insufficient to overcome the force of the rain and wind, and it was still a bucking bronco. A wind gust now hit us from below, quickly raising us up a foot and requiring Blocker to instinctively reduce lift. Then the wind shifted again and the nose moved left by itself as we found ourselves aligned with the fence again. Seeing his opportunity, Blocker wasted no time. He quickly pushed the stick to the right and the Huey jumped away from the fence, out over clear pavement where he could set it down. He was sweating buckets and so was I, and the crew chief mumbled, "I thought we were dead," an observation with which we all silently agreed. A few minutes after Blocker shut the Huey down the rainstorm passed and the sun came out, returning us to a tropical steam bath of oppressive humidity. Some Cambodian civilians appeared and unloaded the fuel drums, and after the crew chief filled the Huey with the hand pump, we filled up on C rations and cigarettes.

Thirty minutes later COL Kenzel returned from meetings in the terminal and told us the Cambodians had prepared dinner for us. We were stuffed from the C rations and told Kenzel there was no need for another meal, but he explained that we were duty bound to go eat what they had prepared because it was the diplomatic thing to do. A couple of minutes later we piled into a couple of vans and were delivered at the steps leading into the terminal, expecting we would be given a hamburger and Coke.

We walked up the steps and entered the terminal through a set of double glass doors. The place was air conditioned and completely empty, and coming in from the humid early evening air made it feel frigid. The terminal was exquisitely finished with marble floors, lots of palms, banana trees and Norfolk pines, and large Oriental carpets with vibrant colors and detailed designs. There were statues of Oriental figures including a large marble figure of an elephant, and the three or four Cambodians who were there to serve us were dressed from head to toe in bright white serving suits. The whole place was immaculate and a feast for the eyes, far cleaner than anything we were used to in Can Tho.

We were directed to sit on floor mats that had been placed on a large rug on the main floor of the terminal, next to several low tables, Japanese style. Once we were settled the servers began bringing food, and for the next hour there was no let up. What we thought would be hamburgers and Cokes turned out to be a full seven-course meal with fresh shrimp and rice as the entrée, and how we stuffed it down on top of the C rations we had just eaten I don't remember. But we did!

The Cambodians serving us were quiet, friendly and polite in the extreme. They seemed to be completely comfortable around us, even though I remained on edge, suspicious about them and concerned about the security at the airport. The suspicions decreased, however, when they brought out beer and wine after dinner and we proceeded to consume as much as we could.

During dinner we found out that our hosts had prepared a place for the officers to sleep in the terminal, so we went back to the Hueys and got our bags. When we returned the Cambodians insisted on serving more beer and wine, and as time passed and I became used to the surroundings, my suspicions evaporated.

About 11 PM we quit drinking and went into the section of the terminal where we were to sleep. It was a beautiful room with all sorts of silver and pearl goods on display, even more exquisite than the main terminal. I struck up a half hour conversation in French with a guard who was posted just outside our door, a soft spoken young man with manners equal to those of the people who served us at dinner. But by midnight I was bushed and found an empty straw mat to lie down on. The air conditioning was still running at full blast, so I put on a second flight suit over my first to keep warm.

**The Phnom Penh Mission, Day 2**

But two flight suits weren't enough to counteract the air conditioning, so by 5:15 AM I was up and trying to get warm.

Breakfast had been prepared and the enlisted crew members who had slept in the Hueys joined us. They looked rested and probably got more sleep than I did, outdoors in the warmer air. While we were eating I stuffed a small placemat into my pocket as a souvenir. I believe the writing on it said "Khmer Society of Royal Restaurants."

It was a beautiful day without a hint of rain, so after breakfast we cleaned up the Hueys and with time to spare strolled around the airport to see what the Cambodians had in their Air Force. It was a strange mixture of planes ranging from a huge Russian bi-plane and several MIG 17s, to French-made fighters and observation planes. There were multiple DC-4s and a couple of French Alouette helicopters that one of the guards said had been used by Prince Sihanouk before he was ousted from office in March and replaced by Lon Nol.

At about 8:30 a group of secret service agents arrived in a jeep with all sorts of radios and other electronic equipment. They looked and acted like tourists, very friendly and unassuming. But beneath this facade they were all business. At 9:00 our rehearsal began. We cranked the Hueys and hopped them over to the taxiway as if we were waiting for Air Force 2 to land, then took to the air and followed the main highway into downtown Phnom Penh, circling over the city and the palace where the VP would meet with President Nol. Then we headed back to the airport where we shut down and waited.

We parked next to a taxiway, and a small motorcade of civilian vehicles, including a white limousine, pulled up behind us and parked. At 11:15 Air Force 2 appeared, entered the traffic pattern and landed, taxiing to its predetermined location next to our taxiway. It was a stunningly beautiful plane, glossy-bright white and sky blue, with capital letters across the fuselage that said, "UNITED STATES OF AMERICA," and an American flag across the tail. The size and majesty of it, etched into a crystal blue sky, took my breath away.

The exit steps from the plane were being put in place as we cranked up the Hueys again, waiting for the VP to appear. It was still a beautiful morning, and from our perspective the only cloud in the sky was directly above AF 2. Small and oddly shaped, it was riddled with a rainbow of colors from top to bottom, something I'd never seen. I mentioned it to our crew and one of the enlisted men said he had seen something like that when he was just a boy. His grandmother had called it, "fire and brimstone." (It turns out the meteorological name for it is a circumhorizontal arc or ice halo). At the time it seemed like an ominous symbol of the precarious circumstances facing Phnom Penh and the Khmer Republic. Just two days before, 500 enemy soldiers were killed in a big battle only five kilometers from the city.

The motor convoy headed out onto the runway, stopping briefly next to the plane as a few unrecognizable people descended the portable steps. Even though we were only about 50 yards away, I couldn't see the vice president in the crowd. The motor convoy turned and drove past us on the taxiway, pausing briefly next to our Hueys where there was a confusing flurry of activity, and then drove on. (The motor convoy was actually a decoy designed to divert enemy attention from the Hueys. VP Agnew had actually gotten out of the limo and onboard chalk 1 when the motorcade slowed down.)

At the direction of COL Kenzel we lifted to a hover and moved onto the runway, then took off and followed the route downtown as rehearsed. We reached the Royal Palace less than 10 minutes later and circled a couple of times above it while chalk 1 delivered the VP. It landed in the palace complex in front of the palace itself. The complex was composed of several buildings on the west bank of the Mekong River at the point where the Tonle Sap River joined it. From 500 feet up it was nearly impossible to see details of the buildings in the complex, but from our altitude it was a beautiful sight.

As soon as Chalk 1 had joined us again we headed back to the airport where we were treated to another full meal just like the night before, but absent the beer and wine. Then we settled into a two + hour wait. At 3:30 we got the call to "saddle up" and headed back to the palace. We again circled a couple of times while COL Kenzel picked up the VP, and then returned to the airport concurrently with the motor convoy. We hovered to the taxiway where we put the Hueys down, side by side, facing AF 2 which was still on the runway, just the other side of a wide strip of tall, dried grass.

A couple of minutes later the convoy arrived, passing by us to get to the plane. The convoy arrived at the foot of the portable stairs and stopped. The doors of a limousine opened and several people started up the steps, and at that moment secret service agents began to rise up out of the grass, first one here and then one there, all holding machine guns at waist level. A total of a dozen or so agents, dressed in suits and ties, rose up from one end of the grass strip to the other, and slowly backed up toward Air Force 2.

The convoy moved off down the runway, opening a direct line of sight for us to see the portable stairs, and we watched the agents carefully back up the steps until they were all aboard. Finally a gaggle of newsmen with cameras and long hair dashed up the steps and onboard AF 2. The stairs were taken away, the door was closed. Soon we could hear the engines starting, then it began rolling, and then it was gone. Just like that, the mission was over; and by 5:00 we were on our way back to CTA where we quickly resumed our normal daily routines. Little was spoken of our top secret mission to Phnom Penh. It was the last time I was in Cambodia during the war.

Two months later, written Army order # 9826, dated 5 October 1970, awarded the Army Commendation Medal to Leopard, Gregg and me for the mission to Phnom Penh. This unexpected recognition was a bit overdone. We had seen no hostile forces and had taken no fire during the mission, plus we had been treated like royalty by the Cambodians while we were there. What we had experienced was a cake walk compared to the combat missions we normally flew, so I was a bit embarrassed by such recognition. Never-the-less, I was pleased that the vice president had expressed his "personal congratulations."

**No Longer a Work in Process**

By late August, I was bored with the motor pool and its problems. I was restless and needed a new challenge, and based on my experience so far I thought it would be found by spending more time flying. So when I received orders indicating that 1LT Duane Henderson was replacing me as motor officer on the $29^{th}$, I was both pleased and relieved. Now I was free to concentrate on being the best assistant platoon leader for the Boomerang flight that I could be, without ground duties distracting me.

This was quite a change from the risk avoidance behavior I had adopted a few months earlier as a survival strategy. Rather than seeming risky, flying now seemed like a relief from the grinding boredom of duties on the ground. I had found out for myself what Tom Brown had told me about in flight school: flying in combat produced a rush that was irresistible, and now I wanted more of it.

I had accumulated more than 120 hours of combat flight time, not a lot compared to the time accumulated by many other pilots. But in that time I had flown through severe weather, been ambushed and shot down, and had made a bloody medevac. I had seen pilots do things the Huey wasn't designed for to save the lives of comrades, had bullets bounce off my boot and pass through the cockpit within inches of my head, and had learned to endure extremely long hours in the pilot's seat. I had made some bad decisions, but a lot of good ones too, and had become unflinching, efficient and effective. I had become what the situation had demanded of me.

The full impact of that change on both psyche and soul was not comprehensible then, and there is only a vague understanding of it more than 40 years later. What is apparent is this: as of that moment I was no longer afraid, and combat conditions no longer controlled me. From that point on, when I was in the air I was most concerned about the mission and the welfare both of those around me and those I was sent to Vietnam to protect. I was controlling the situation now, but I didn't realize this would not last the rest of my tour.

Along with being unafraid I was becoming confident, and increased responsibility was heading my way. I had already been designated the pilot in charge of the Nighthawk mission; Simon had picked me to be a future air mission commander for the Boomerang flight; and Dave Armstrong told me before he left for the States that I was good enough to be an aircraft commander right now (but I was short by a hundred hours of the 300 that were required). I thought my future would be as an air mission commander, but that notion didn't last long.

**You're Fired!**

Shortly after Major Simon took command of the company, he decided to select one of the company pilots as a permanent copilot, and the unfortunate pilot turned out to be me. After this news started getting around the company, various pilots came up and said things like, "good luck, you'll need it," or "better you than me," or "I wouldn't wish that on anybody," all designed to reflect their gratitude at having been spared the agony of breaking in a green company commander. Why he picked a relatively inexperienced officer like me when he could have selected any pilot in the company was a mystery.

Simon's combat role as Boomerang 6 was command and control (C&C) of all the combat assault missions flown by the 191st. From his "perch" at 1,500 feet above the ground, he directed the Boomerang flight and coordinated support resources and combat details with his counterparts in the infantry units we were supporting on the ground. As C&C, there was always a fifth person in the Huey who communicated with the ground units. He was usually referred to as the "backseat" because he sat in a temporary seat centered behind the pilots. Backseats were usually captains or majors, and if we were supporting ARVN troops they were Vietnamese officers.

By this time I had learned specific, unwritten rules that the pilots in the flight observed carefully. One of them was to stay off the radio and refrain from talking while the flight was on final approach to an LZ. This was primarily a safety precaution in case there were warnings of enemy fire, or specific orders needed to be given to the flight due to unexpected complications. Since these rules were unwritten, the only way to know them was through the experience of flying with the flight.

In late August we were flying with a rotund, talkative major who was serving as that day's backseat. He and Simon were making small talk about things unrelated to the mission while I was listening carefully to the talk between the pilots in the flight. This was one of the company's first missions after all the experienced ACs had left earlier in the month, and the new ACs were having a rough time. When the flight started its descent to the LZ, Simon and the fat Backseat continued their chatter as if they were in the corner café, making it nearly impossible to hear the talk within the flight. This went on ceaselessly as the two majors continued their ignorant behavior, and my blood began to bubble at their disregard for the flight's safety.

Fed up with being unable to hear the radio discussion between the Boomerang pilots, I gruffly said to Simon and his new-found friend, "Will you please shut up! We have a rule in this company to remain quiet while the flight is on final and with you two talking I can't hear anything!" I went on to explain the importance of the policy and how everyone was obligated to observe it; and to their credit they clammed up immediately.

I think both the Backseat and Simon were taken aback that a first lieutenant would talk to them that abruptly, and I suppose I could have handled it more tactfully. But the selfishness and disregard for the Boomerang crews they displayed triggered an anger that demanded to be heard. Their behavior was yet another sign of the disrespect that I had seen over and over on active duty, but this time it was disrespect for the enemy's capabilities. It was only a few weeks since Mahan and I had been shot down, and with that experience still fresh in my mind I was intolerant of the ignorance and disrespect these two senior officers had shown.

The rest of the missions that day proceeded with no problems and there were no enemy encounters. When we returned to Can Tho late in the afternoon, I parked the Huey in its revetment, filled out the logbook and watched from the pilot's seat as the tanker truck arrived and the crew chief refueled the aircraft.

When I stepped out of the Huey, Simon confronted me about the outburst earlier in the day, and in a modest and respectful manner said my behavior had been unacceptable. I don't remember what else was said, but a day later I was back flying the Nighthawk mission, which was OK with me. My days as the commander's copilot were over and I considered it a good development.

After my dismissal, a junior warrant officer was chosen as the Boomerang 6 copilot, and during the remainder of Simon's command of the 191$^{st}$ we managed to stay out of each other's way. Except for an occasional awards ceremony or a conversation about work, I saw him very little and once his tour was over, I never saw him again.

In 2011, forty-one years after my tour ended, I received an unexpected telephone call from Joe Babbitt, a Bounty Hunter pilot with the 191st in 1970 who was trying to locate former members of the company. He briefly ran through people he had located and talked to, one of whom was Blake Simon.  Joe said he had a brief conversation with him during which he remarked that he was surprised to get Joe's call, because he never thought he would be hearing from anyone in the 191st!

# Chapter 5 - SEPTEMBER 1970

I awoke one morning and it was September. The calendar had changed, but nothing else had. There were still combat assaults, ash & trash, and night missions to fly, and it was back to Rach Gia for me. This place, where I had seen some seriously big things happen, was where numerous small things happened as well, like the following two stories show.

| *Fire and Rain* - James Taylor |
| --- |

### Taking a Hit

One early morning I was flying copilot on the lightship with Gordon Clayton. We were at low altitude and the crew chief was firing grenades out the cargo door from an M-40 grenade launcher. I turned to watch him fire and to see where the grenades were hitting, when one of them exploded prematurely about 50 feet away from the Huey. I felt a sharp sting in my forehead just above my right eye, and with my finger I could feel something embedded there, which turned out to be a small piece of shrapnel from the grenade. After that little incident, I almost always wore my helmet's plexiglass visor down to protect my eyes. There is still a scar over my eye where the shrapnel hit.

### How Much Does a Huey Cost?

Another incident occurred early one evening when we were loading pyrotechnic flares into the flare tubs that were mounted next to the cargo doors on each side of the lightship. Aircraft flares were light weight and once lit they burned with an intensely hot chemical flame, usually white phosphorous.

I was helping the crew chief load and arm the flares, each of which were contained in a metal tube about six inches in diameter and three feet long. Each flare had to be armed before it was loaded into the tub by turning an arming switch on the end of the flare where a metal lanyard (wire) extended about three feet from the end of the flare. The lanyard would then be clipped to a metal ring in the floor of the Huey, and when it was removed from the tub and thrown out at altitude, the lanyard would pull out which started the flare burning and deployed a silk parachute at the same time. The flare would then descend slowly, lighting the countryside below.

The crew chief and I were talking about unimportant trivia while we armed and loaded the flares. As we worked and talked, I accidentally stood on the lanyard of an armed flare and when I picked it up to toss it into the tub the lanyard pulled out of the flare. There was a "pop" and then a hissing noise and smoke, and the metal tube blew completely off the flare and parachute that were neatly packed inside. Thinking that the flare was going to ignite, I yelled at the crew chief to run because the flare was going to burn. We took off in the same direction and ran about 50 feet or so to the rear of the Huey where we waited for the flare to burn.

My mind was racing with questions: how much damage could a flare, which had an intensely hot flame, do to a Huey? Would the flare ignite the fuel in the Huey and burn it to a crisp? If that happened, how long would it take me to pay for the loss of a $250,000 helicopter? Would I be court-martialed and sent to jail?

We waited and watched for several agonizing minutes ... but nothing happened! Cautiously approaching the Huey we could see the flare itself lying next to the white parachute which was still in the shape of the tube it had been in. The tube was lying about 10 feet to the front of the Huey. I went and picked it up, then tried to repack it with the flare and parachute, but they wouldn't fit. (Once you open a can of worms it's impossible to get the worms back in!) Nothing had happened, no one was hurt and there was no destruction. Only our nerves had been jangled.

I later learned that a flare would not start burning until its parachute had opened in the air and the resulting jerk from the weight of the flare below it caused it to ignite. Another lesson learned the hard way, but not as hard as it might have been!

--- Ω ---

Saturday night, 5 September, I flew the daily Phantom Twilight (PT) and Can Tho Firefly (CTFF) missions designed to protect the airfield and Eakin Compound in downtown Can Tho from attack. PT involved checking the barbed wire barrier around the perimeter of the airfield from the air during daylight to be sure there were no holes or booby traps in it that would permit the VC to have access to the airfield after dark. After checking the wire we became the C&C for two Cobra gunships that were given free-fire areas near the airfield where they fired on targets of opportunity, like suspicious Vietnamese sampans.

CTFF involved a C&C Huey and a lightship that patrolled the perimeter looking for VC who might be trying to break through the wire, or set up positions to mortar the airfield. There were two primary times when the CTFF would be in the air for two hours each, 11:30 PM and 4:30 AM. The rest of the time the crews would be on standby in case there was an unexpected attack, either at the airfield or at Eakin. Most of the time these missions were boring, but they were necessary to provide security.

--- Ω ---

On Wednesday, 9 September, I was the airfield officer of the day (AOD), in charge of ground security for the airfield that night. Working from the $13^{th}$ Battalion Operations Center on the north side of the airfield, I drove the entire perimeter checking each of the numerous two-man bunkers that guarded the perimeter wire with M-60 machine guns, rifles, grenades and Claymore antipersonnel mines. Although I had flown around it just four nights previous, it was the first time I'd driven around the entire perimeter. I had not realized just how big the airfield was until then.

At 4:00 the next morning, the sergeant of the guard and I heard an explosion that came from the south side of the airfield. It was immediately reported to the military police, who investigated and reported that someone threw a live grenade at the door of the commander of one of the Chinook companies across the runway. We never found out if anyone was caught. That particular fragging incident would not be the last one at CTA.

**Stupid is as Stupid Does**

The nicest thing about ash & trash missions was the break from the dangers of combat assaults and the variety of things we experienced. One mission involved delivering some confidential documents to an outpost in the northwest corner of the delta. This particular outpost had a landing pad about 75' square that was wedged between steep mountains on three sides, leaving only one way in and out. I was the pilot in command that day, flying with a warrant officer copilot who had been recently transferred into the company from another unit.

We reached the outpost after a 40 minute flight from Can Tho in clear and sunny weather, and the copilot made a successful landing through turbulent wind drafts that danced around the mountains in hot, sunny weather. I had radioed the outpost to tell them we were arriving with a confidential package and an enlisted man met us when we touched down. After the drop-off, the copilot brought us to a hover, then keyed the intercom and asked if I wanted to take off backwards. I asked what he meant by that. He replied that he would start moving the Huey backwards, and after reaching translational lift (at 15 knots) he would turn 180 degrees and continue the takeoff moving forward.

I was no prude when it came to taking reasonable risks. However, each day in combat brought a load of risks, many that were unavoidable, and I realized from the incident of turning the Huey's tail the wrong way with Stiles just how easily that could happen. This proposal was not only unnecessary, it was ridiculously poor judgment by this young warrant officer. After a moment to recover from my astonishment, I told him to take off normally. He seemed genuinely disappointed, but followed my instructions.

I knew that unnecessary risks should not be taken if one was to survive in Vietnam, a place where people were gunning for you every day. This experienced copilot should have known better than to willingly compromise our safety, and he immediately went to the top of my "most unpopular pilots" list. When we got back to CTA, I saw to it that I never flew with him again.

**A Rare Day**

We were in the free-fire zone on the Nighthawk mission. It was just after dawn and the only thing keeping us awake was the cool fresh air that blew into the cockpit.

Below us the Bounty Hunters were working over an area south of the U Minh (*yew min*) Forest below Rach Gia, machine gunning tree lines and rocketing whatever looked like it might be a bunker. Fifteen hundred feet below us was a small grassy clearing with a thatch-roofed hut on the edge of a pond of black water about a hundred feet across that dominated the center of the clearing.

A machine gun was turned on the hut forcing someone in black pajamas out the front door who dove into the pond and began swimming across. The gun pilot radioed that the swimmer had something on his back that looked like a radio. The door gunner turned his sights on the swimmer and started to squeeze the trigger of his M-60. Uncertain about the target's identity, the gun pilot told him to hold off so he could get a better look. The Huey circled down over the pond in a tight spiral. Fifty feet over the swimmer he realized it was a woman, and what looked like a radio was actually a small child clinging to her back. The situation met the minimum requirements of the Rules of Engagement. She was a legal target.

The gun pilot radioed he didn't want to fire on a woman with a baby, even if she was VC. We were not baby killers. We turned and flew away.

Adults save babies; rarely is it the other way around.

--- Ω ---

One of the hardest things to deal with during any night mission was sleepiness. If there was no action going on, sleep was never far away and there were many occasions when the crew would nod off. With the FM radio tuned to AFVN at night, we could hear the smooth voice of Dean Martin or the cascading strings of Mantovani, mesmerizing, drowsy sounds to an ever-weary crew. Al referred to the unceasing call for sleep as the "rack monster" (or bed monster).

If I wasn't too sleepy, I'd fly while the rest of the crew caught 40 winks (OK, maybe only 39). But, occasionally my eyelids were just too heavy to hold up and I, too, would drift off, snapping awake with a jolt a few seconds later to find the Huey drifting into a gently descending bank to the left or right. This would usually scare me into being wide awake for the rest of the night.

Nothing ever happened from grabbing 3 or 4 winks like that. But I never would have admitted to being asleep at the stick to anyone else.

**Oklahoma Attitude**

I met Allen Hoppes when we were students at Hunter Army Airfield in class 69-50. I knew who he was, but I didn't know anything about him until we had a committee meeting to plan our graduation party. We were on the committee together, and during one discussion about how we were going to get the dinner tables and chairs set up, Al volunteered himself and his wife to take care of it. It was obvious from this example that Al was a team player, humble enough to serve others in a menial task while showing initiative and leadership. I was impressed by his character.

Al was good looking, about 6' tall with a medium build and dark hair. He was smart too, and had a degree in Engineering from the University of Oklahoma. His most profound passion was Sooner football, but being equally passionate about Ohio State football we could not avoid emphatic "discussions" about which school had the best football program. We each had valid claims to make: Oklahoma had the longest college football winning streak in the 1950s (a record that still stands) and also the 1969 Heisman Trophy winner in fullback Steve Owens. Ohio State had won 5 national championships including 1968 when they beat USC and Heisman Trophy winner, O.J. Simpson, in the Rose Bowl. Our discussions on football usually resulted in a draw, both of us being careful to maintain the friendship above winning the debate.

Other pilots in the company, bored sometimes as the result of too little flying time, occasionally wandered into "discussions" with us about the football programs at their alma maters. Many were from small colleges or Ivy League schools whose football programs could never match behemoths like Oklahoma or Ohio State. Invariably they would retire from such discussions with their tails between their legs, beaten to a pulp from the tirade of statistics that included the number of national championships and Heisman Trophies won by Buckeye and Sooner players. Our verbal drubbings were administered mercilessly and without discrimination.

Most pilots wilted under our verbal barrages, but not Rick Mahan. Rick had attended Sam Houston State, but was a Texas

Longhorns fan through and through. He could boast about college football as well we could, standing toe to toe with Al about Red River shootouts and Texas-OU weekends, aspects of the rivalry between the states of Texas and Oklahoma that I didn't understand at the time. His ammunition was the Longhorns' national championship in 1969 that resulted from an 11 and 0 record that season. We had met our match!

Al liked jokes, including practical ones, and somewhere along the line he was given the nickname "Uncle Al, the kiddies' pal," based on a character of that name played by Alan Sues on a TV comedy called *Laugh-In* that was popular at the time. Al was a good one to joke with because he was patient and even-tempered, never angry, and whether kidding or serious, everyone liked him. He was generous and understanding, relatively non-judgmental, an all-around good guy with a sincere love for the American southwest. He sometimes referred to himself as an "Okie from Muskogee," which was a popular song recorded live by Merle Haggard in Muskogee, Oklahoma in 1969.

As the company administration officer, Al spent most days at the company headquarters supervising the paper flow through the company office. On flying days he flew with the 2nd Platoon, the Green Deltas, ferrying generals, congressmen, diplomats and other VIPs around the delta to visit provincial compounds or outposts. Such duty was devoid of much exposure to real combat, but dealing with the egos of the VIPs often required great patience. It wouldn't have been my cup of tea, but it seemed a perfect fit for Al's patience and diplomacy.

## CYA

The Army has always been known for its many bureaucratic idiosyncrasies, some of which existed due to real needs, and others that had lost their true purposes long before, but resisted being done away with. One feature of this culture was the proliferation of duty titles that connoted mundane responsibilities which were always delegated to lower ranking officers. Such delegation was formalized by the company commander with written orders that permitted him to duck responsibility, or "cover your ass" (CYA), in case something went wrong. It was a way to designate a scapegoat in advance, and guarantee there would be some lower ranking officer to blame if something went wrong. It was an institutionalized practice that guaranteed feces would always flow downhill, away from the commander.

Just to illustrate how ridiculous the situation had become by 1970, here is a list of formal duties that Al held while he was the admin officer: Forms Management Officer, Unit Historical Officer, Legal Assistant, Postal Officer, Publications Officer, Records Management Officer, Ration Control Card Officer, Witnessing Officer, Unit Fund Custodian, AER Officer (whatever that was), C Day Conversion Officer, and R&R Officer. In Late August all these "critical mission duties" were reassigned to WO1 Rory Candelbach, Al's replacement as admin officer. With a quick swipe of his pen on the orders, the CO assured that feces would continue to flow downhill to his newly designated scapegoat.

All junior officers were candidates for scapegoat duty. On October 20th, 191st Unit Orders # 79 were issued by Major

Simon appointing John R. Barth the censorship officer for the company. I was not told this appointment was coming; it just appeared out of the blue. Nor was I briefed by anyone as to what the duties of the censorship officer were. In fact, I never did anything as censorship officer: I never read and edited anyone's mail, I inspected no outgoing packages, and I questioned no one about the contents of their communications, written or otherwise. In fact, nothing at all happened. But the CO had covered his ass.

**Two Steps Forward, One Step Back**

Late in September the first written evaluation of my performance as an officer in combat was completed by Ron McConkey, the 1st Platoon commander. His rating was complimentary and included a positive remark about the progress I had made in the motor pool operations. MAJ Simon declined to make any comment on the basis of having had too little time to observe my performance to be objective. (In reality, objective evaluations were almost never possible because human bias could not be eliminated from the rating process). Getting written feedback on performance in combat didn't occur very often. However, on a daily basis it boiled down to survival and mission effectiveness, and I felt good about accomplishing both. This first combat valuation was important to me, and I found McConkey's comments encouraging.

However, none of it jibed with finding out a few days later that I was under investigation for the loss of unexpendable tools that were missing from the motor pool. Unexpendable tools were the expensive ones that could not be replaced just by ordering replacements; they had to be accounted for, and if any turned up missing the officer in charge could be found responsible and have to replace them from his own pay.

The first indication I had of the investigation was a rumor out of the CO's office. A few days later when I was alone on the flight line I was confronted by the investigating officer who wanted a statement from me about the loss of the tools. I gave him my side of the story, that the tools had never been in the motor pool during my assignment as motor officer, and that I had no idea where they went. As far as I could tell, they were not there when I arrived.

This weasely looking captain, who seemed suspicious of everything I said, was not satisfied with my statement and openly accused me of lying. He declared that he would see to it that I was court-martialed for dereliction of duty, a comment that greatly distressed me. I believed I was doing the best I could with what I had, and his attitude seemed unfair and inappropriate for the circumstances. With great difficulty I managed to restrain myself and made no reply, and he turned abruptly and walked off leaving me alone again to figure out what might happen next. But as it turned out, nothing happened next. A few weeks after that confrontation a report was issued saying that the missing tools in fact had been missing when I took over the motor pool, and had been missing since the 191$^{st}$ had moved from Dong Tam to Can Tho in 1969.

I realized I should have inventoried all the tools in the motor pool when I arrived as motor officer and reported the missing tools then, but I hadn't. It was a fundamental error on my part, but not a felony (or even a solid misdemeanor for that matter). However, because of its timing this incident took some of the luster off the positive efficiency report that Captain McConkey had given me. It was a good example of taking "two steps forward and one step back" as the saying goes.

This minor flap resulted in personal humiliation, even though it appeared to have no material effect on me or my situation. I had made a mistake, not one that would prove lethal like a bad combat decision might, but one that raised a question about my own decision making. I was definitely not comfortable trusting the decision making capacity of some of the others around me, but if I couldn't trust my own judgment whose could I trust?

--- Ω ---

Whenever we were at cruising altitude in our Hueys, we flew at 1,500 feet above the ground, an altitude that was beyond the effective range of small arms fire, like AK-47s, the Russian-made rifle used by the VC. We had little concern for longer range, larger munitions like radar-controlled .51 caliber machine guns, because such large weapons were not available to the enemy in the area south of Saigon, so we were told.

Early the evening of 13 September, I was flying in the right seat with Rick Mahan south of Saigon in the III Corps area. We had been on a Delta Courier mission all day and were on our way back to Can Tho. Rick was flying and I was daydreaming, looking out the right cockpit door window as the dusk slowly settled into darkness.

As I stared out the window, what looked like burning balls about the size of softballs began rising slowly beside us, maybe 25-30 feet from my door. Some were green, others red, gold and silver in color, all of them sparkling like a Fourth of July sparkler we used to burn as kids. They were mesmerizingly beautiful as they slowly rose past our Huey and gradually out of sight, and I was fascinated by this unexpected light show, a phenomenon never seen before. A moment later I realized something was wrong and turned to Rick and said, "Hey, look," pointing out my door. "What's all that?" I asked him. Rick turned and looked past me out the door, and as he did we heard a high pitched beep come from one of the radios. All at once it occurred to us that we were being tracked and shot at by a radar-controlled, large caliber gun.

We had been told in flight school about radios being able to pick up radar beams from enemy guns, but since no one ever ran into that stuff in the Delta we were not alert to it. What we had seen were the gun's tracer rounds, and the beep we heard was the radar beam hitting our Huey as it swept past us. As the reality of the situation sank in, Rick immediately pulled pitch and climbed to 2,500 feet to get out of range of whatever was shooting at us.

When we were finally clear we again looked at each other, wiped the sweat off our faces and privately thanked God for escaping what could have been disaster. The crew chief and door gunner had been completely quiet during the episode, and I'm not sure they weren't asleep throughout the whole thing. How a notoriously accurate radar-controlled gun had missed us was a mystery.

--- Ω ---

Combat activity throughout much of the area south of Saigon had been tapering off since the invasion of Cambodia in May, except for the area south and west of Ca Mau. While Mahan and I were flying Delta Courier on the 13th, the pilots of the 162nd and one of the local cavalry troops were getting their tails shot off in combat assaults south of Ca Mau. The following day, CWO Brad Crandall and I flew lead position in the Boomerang flight inserting ARVN troops into the same area. I'd flown with the flight a number of times by then, but it was the first time I'd flown in the lead position, Chalk 1, and it turned out to be a great learning experience.

Brad Crandall was a native of the state of Wyoming who trained in Flight Class 69-29 and graduated about 20 weeks ahead of me. He was an experienced AC, friendly but quiet, about 5'8" tall, and he knew what he was doing in the cockpit. While he flew and communicated with the rest of the flight, I had my hands full navigating us to the LZ using a map and grid coordinates. It was very different from flying any other position in the flight where we simply followed the leader.

Flying in Chalks 2-5 in a CA required the ability to handle the Huey effectively, communicate well with the lead pilots and the crew, and manage the ground troops who were the "cargo." With experience these limited responsibilities became routine, almost second nature, even when the flight was taking fire. But flying lead was a whole new ball game that involved thinking about, and for, the entire flight. It especially required staying ahead of the situation by anticipating what might go wrong, and knowing what to do if unexpected events developed and "things went to shit" as the pilots would say.

Half the time I had my head buried in the map trying to find the landing zones on the ground from the map coordinates we had been given by C&C. The rest of the time was spent giving the other pilots in the flight instructions on how to approach and leave the LZs, communicating with C&C, and flying smoothly so the rest of the flight could follow, all while handling the regular duties as one of the Hueys lifting troops from the PZ to the LZ. The time went fast and I made no significant mistakes. The LZs were cold; we took no fire and no casualties. It was a good day for the Boomerangs.

The following day several units operating in the exact same area we had been in got the hell shot out of them. The Navy Seawolves out of Rach Gia had four Hueys shot down, losing a pilot and one crew chief in the process. A medevac was lost trying to pick up the wounded from the downed crews, and the 336th Warriors from Soc Trang also lost four Hueys. Altogether 12 aircraft were lost that day, including a Huey that was shot down in VC Lake about 10 miles west of Ca Mau and could not be recovered. It was blown up with rockets to keep it out of VC hands.

It was another amazing example of the arbitrariness of battle that had delivered the Boomerangs from harm's way on Monday, while other units were shredded the day before and the day after in the same area. A Boomerang always comes back.

--- Ω ---

About 40 miles up the west coast of the delta from Rach Gia and snuggled up against the border of Cambodia was the village of Ha Tien. It was an old village, settled by ethnic Chinese who supported the Ming Dynasty in China about the middle of the 17th century, when the Nguyen Lords ruled southern Vietnam. The Chinese settlers built markets there that created a commercial center, and over time Vietnamese moved in and the village eventually became part of Vietnam. Today, Ha Tien is a small city of 40,000, known for its beautiful beaches and the hydrofoil that runs to Phu Quoc Island twice a day. In 1970, it was known to us as the village near the cement plant.

Late in the day on 15 September, we flew a Vietnamese technician to the cement plant to repair the generators that powered the plant. It was after dark and the night was clear and beautiful, with a few wispy clouds floating softly overhead. On the return trip to CTA thunderstorms threatened the area from the west, but to the east a full moon lit the landscape as far as the eye could see and reflected brightly off the rice paddies below us.

Such beautiful nights in the air belied what was probably happening on the ground 50 miles to our south where the VC controlled the night. I didn't know which pilots from our company were flying Rach Gia Nighthawk, but thinking about it brought back memories of the night in June when we were ambushed at Rach Soi, memories of hitting the runway hard, and leeches, and Porky McCarthy. They were memories I wished I didn't have, memories of an incident that had forever changed my internal landscape and made one fact as crystal clear as the moonlight: there were always people down there who wanted us dead.

After returning to CTA, I spent the rest of the night helping Henry Solomon recreate missing historical records for a Huey that was being retrograded the next day. Sol was always in good humor and it was a pleasure to be around him, but we didn't get done until 5:00 AM and by then I was beat.

Later on the 16th, I again flew the Phantom Twilight mission around the airfield. Compared to combat assaults and Nighthawk, it was like having a day off.

--- Ω ---

In addition to the United States, other nations provided military support and personnel to South Vietnam during the war. The list included Australia, South Korea, Thailand, Philippines, New Zealand, Cambodia and Laos. We rarely heard anything about how the troops from those other countries operated, except the South Koreans who had a reputation for brutality and didn't take prisoners (they killed them instead).

We also heard now and then about Australian helicopter units, of which there were two: the 9th Squadron of the Royal Australian Air Force, and the Experimental Military Unit, or EMUs for short. The EMUs were an assault helicopter company (technically the U.S. 135th AHC) that was composed of two-thirds U.S. pilots and one-third Australian Navy pilots operating out of Dong Tam, northeast of Can Tho. The Australian EMUs had a reputation for being down to earth, cool under pressure, effective at what they did, and fun loving party-goers.

The evening of 17 September there was a show at the O Club that had two Australian girls in it. The show was not memorable, and the girls were not particularly attractive, but they were intriguing with their party girl personalities and unusual English accents, described by one writer as "a kind of fossilized Cockney." Our gun pilots were spellbound by the girls and invited them to the Bounty Hunter lounge for a party after the show. Al and I were invited to come along, an unusual opportunity since the Bounty Hunters almost never allowed non-gun pilots into their midst, or into their lounge. I opted not to go and to get some rest instead. But Al went and graciously shared his Okie humor with the Australian guests.

The Bounty Hunters were all nice fellows one-on-one, and I appreciated that they risked their necks to protect me and the Boomerang flight. In fact, I may have owed my life to the Bounty Hunter pilots who were with us the night of 26 June at Rach Soi. They were good at what they did, and I was grateful for that. In general, gunship pilots' work required a special combination of detachment and fierce cold-bloodedness. I didn't understand what it took to produce that combination, but I knew I didn't have it.

**The Wingnuts**

The Maintenance Platoon leader, known as Wingnut 6, was an unusual man. CPT Don Jefferson was enigmatically distant and personable at the same time. Like Ron McConkey and Connie Leopard, he had been an enlisted man who went through OCS, and his enlisted experience gave him a sergeant's moxie and force of personality. When he said to do something, people did it with no hesitation.

At 5'10' tall, with a perfect black brush cut, a slender build and a casual manner, Jefferson was hard to figure out. He could be easy-going one minute and demanding the next, or by-the-book in one instance and completely ad hoc in another. While he appeared to be secretive and unpredictable, I found him to be approachable, unflappable and reasonable on issues regarding work.

One of his not-so-secret secrets involved a girlfriend, a round-eyed Red Cross worker he'd arranged to get assigned to Binh Thuy. I asked him one day why he would choose to bring a female he cared about into a war zone where she was exposed to physical danger. His eyes narrowed to slits when I asked the question, and with pursed lips he told me to mind my own business, which is exactly what I did.

There were other unique personalities in the Wingnut Platoon. One was 1LT Roger Murdock, who possessed a deep Southern drawl and the personality of a snake. He was arrogant and critical and I disliked him intensely because he had nothing to say that was good about anybody, so far as I could tell.

Another was CPT Will Moss, a mild mannered pilot who was the assistant platoon leader, Wingnut 5. He was about six feet tall and slender, with thinning brown hair and a mustache, and bounced when he walked like he had springs in his heels. Will was not exactly unfriendly, but he rarely smiled and was always slightly aloof. His equivocal nature made his moods hard to read, and in that way he resembled Jefferson, with whom he got along well.

A fourth pilot in the platoon was WO Dexter Woodley from northwestern Minnesota, who was a maintenance test pilot. He had a strong, square jaw that gave him a profile that could have been used as a model for the prow of a sailing ship. Dex was about 5'6" tall with all the physical attributes of a wrestler. He was consistent and dependable, someone I would have to lean on heavily in the coming months.

There were about 40 mechanics in the platoon and a platoon sergeant (a very nice SFC whose name I can't remember) all working within a hangar that could hold four Hueys side by side, but was shared with the 162nd. A civilian sheet metal technician under contract to the Army handled all of our sheet metal repairs in a small shop at the back of the hangar. The maintenance office was above a workshop area on the left side of the hangar, about

10 feet above the hangar floor, and that's where "Jefferson's Board" was.

Mounted on the hangar wall in the middle of the office was a 4' x 4' piece of white painted plywood that Jefferson used to keep track of how close each Huey in the company was to its mandatory 100 hour periodic examination (PE for short). Each Huey was represented by a small piece of sheet metal cut in the silhouette of a helicopter which was stuck onto the board with a piece of Velcro. The pieces were moved daily from the upper left corner of the board toward the lower right corner, based on the number of hours it flew. When it hit the bottom right corner of the board, it was pulled into the hangar for maintenance.

I admired Jefferson's Board. It was a beautiful example of the simple ingenuity that abounded in the Army. It was durable, cheap to make, fabricated from materials readily at hand, had no moving parts of its own, and could be understood by a third grader. In short, it was effective, so effective in fact that pilots from around the area would visit the hangar and climb the steps to the office to view the board with near reverence. It was a marvel of management effectiveness and no one touched it except Jefferson, the Master of the Board. Only he could "fly Hueys into PE" as he called it.

But by late September the Wingnut Platoon was poised for significant personnel changes. Jefferson was "short" (i.e. had very little time left on his tour) and would soon no longer be there to fly the Hueys into PE. Bad Attitude Murdock and Will Moss were even shorter. So, in a platoon leaders' meeting with the CO on the 26th, Jefferson was asked who he wanted assigned to the platoon as a replacement. He replied "LT Barth," and the

CO agreed to move me to the 4th platoon. The deal was sealed by formal orders that were issued on the 29th and I was a Wingnut!

--- Ω ---

Helicopter maintenance officers were supposed to be graduates of the Aviation Maintenance Officers Course (AMOC) to be qualified for maintenance work. I had not attended AMOC, but in my new role I would be a maintenance technician and test pilot that involved duties that could be learned on the job. I was leaving the role as assistant platoon leader of the Boomerang (or 1st) Platoon, for a lesser position in the 4th Platoon. But I had nothing to say about it, and the reassignment offered new learning that I found stimulating. Coincidentally, Al was reassigned at the same time to take my place as Boomerang 15.

I was no longer a slick pilot, but a maintenance test pilot with the call sign of Wingnut 3. In all honesty I had mixed feelings about the change. I was glad to be leaving the slick platoon and to be out of the combat assault business, thus leaving behind the work of being a regular target for the VC. I was also happy to say goodbye to the Nighthawk missions at Rach Gia where I had seen some rough times. But I had learned a lot from people I considered top notch pilots in the Platoon, like Armstrong, Cotton, Mahan and Clayton, and I would miss flying with them. With all there was to learn as a Wingnut, the new assignment promised to be an adventure.

--- Ω ---

The 29th of September also marked the end of a formal investigation I had been given two weeks before to find out what had happened to about 40 machine guns and a number of M-16s that were missing from the company arsenal. All firearms, mortars, canons and the like were accounted for by individual serial numbers that were recorded in property books. Those books were the official records showing what was assigned to a company. The guns I was investigating had come up missing during a recent physical inventory of all the assets in the arms room at the back of the supply shack.

Firearms were not only expensive, they were the most important asset needed to execute the war and had to be accounted for at all times. There was genuine concern that the missing items may have been pilfered and sold on the black market. But more importantly to the CO, missing weapons could bring his career to an abrupt halt. So Simon announced to his staff that he wanted the most objective officer in the company to investigate the matter, and I got the job. Being alone on this assignment and having Simon's gonads in my hand, figuratively speaking, I enjoyed being in this position of power.

There was little flying time for me during those two weeks, due to the considerable research through company records that was required. A complete physical inventory showed that for every serial number that was missing there was an existing serial number in the inventory that was not on the property list. This led to the assumption that weapons had been turned in for repair and replaced with new serial numbered weapons that had not been recorded in the property books. A review of receipts for the trade-ins revealed that was indeed the case. I finished the research and the written report and turned it over to MAJ Simon who was highly relieved to learn there had been no hanky panky, and he would not be staring down a court martial board.

This “objective” investigator had decided not to squeeze Simon’s private parts.

# Chapter 6 - OCTOBER 1970

## The Flavor of Maintenance

Helicopter maintenance operations was a 24/7 business supposedly divided into three daily shifts. Always short of mechanics, however, a primary challenge was staffing, so our mechanics often worked 12 or 15 hours at a crack. On top of regular duties, they were assigned to perimeter guard duty every few days, so they were perpetually tired. We were concerned that the physical demands could result in critical mistakes, but there was never any direct evidence of that.

The other major challenge was having all the repair parts needed to maintain 35 Hueys. With each chopper having more than 2,000 moving parts and an unknown number of non-moving ones, it was a constant challenge that sometimes required cannibalizing parts from Hueys in the hangar to keep Hueys on the flight line mission-ready. Ordered parts were shipped from Long Binh nearly every day and the flow was continuous, but unreliable. It was important to have a good working relationship with the maintenance shops of other companies, like the 162nd, so parts could be borrowed and then returned when they finally did arrive. It was a "you scratch my back and I'll scratch yours" relationship.

Occasionally, backordered parts that could not be found locally and were urgently needed had to be picked up at Long Binh supply operations near Saigon. Sometimes those parts were too bulky and heavy to fit inside a Huey, like a jet engine, and had to be transported by suspending them below the Huey with a sling attached to the belly hook, and flying back to Can Tho with a crate dangling 15 feet below the helicopter.

Flying with a sling load was a ticklish business, because dangling cargo suspended by a strap tended to oscillate in flight like the pendulum of a clock. Sometimes it would oscillate fore and aft, sometimes left and right. And sometimes it would swing in a circle that would increase in size during the flight. But no matter which direction the oscillation took, it affected the control of the Huey in an often unsettling way as the cargo pulled the Huey in the direction of the oscillation. Often the pilot would feel as if the cargo was flying the Huey, not the other way around.

Tail booms were also challenging for us to transport because they were about 20 feet long, rigid, and light. In such a case the tail boom would be slid into the cargo compartment and strapped down to the floor leaving 7 or 8 feet sticking out each side of the Huey. I transported tail booms and engines to Can Tho, always with anxiety, hoping that I didn't encounter another radar-controlled .51 caliber VC gun during the flight!

With my focus totally on maintenance I was quickly learning a lot about how the Huey worked. As a test pilot, most of my work was during the nighttime hours, similar to the Nighthawk mission I had flown many times before. However, Nighthawk was always interspersed with daytime CAs or ash & trash missions so it was hard to have any kind of reliable sleep pattern. As Wingnut 3, I was on duty every night and that meant sleeping during the hottest hours of the day, which was challenging due to the heat and noise from daily operations. But the new schedule seemed to make the time go faster, and working nights offered opportunities for practical jokes, such as this one recorded in a letter dated 6 October 1970:

> ***Just got off work a little while ago. I'm really beginning to enjoy working at night. I'm my own boss and no one is around to bother me. Last night Dex Woodley and I went out and caught a bunch of minnows in a glass jar out of the swamp between our revetments and the runway. The rain has brought the swamp up over its banks and the fish swim around in the shallow water-- easy to catch. We left them in the jar yesterday morning on the Admin Officer's desk in the orderly room. Left a note saying it was a fuel sample from 584 (a M model that crashed the other night because of fuel contamination).***

**Vinh**

Often Al and I would say hello to each other in passing at 6:00 or 7:00 in the morning, as I arrived at the hootch when he was leaving for the flight line for a day mission. Sometimes, when I was too wound up to sleep, I would take off my flight suit to cool off and write a letter to my wife. On more than one occasion our hootch maid, Vinh, would come in to clean while I was sitting in my skivvies. But that didn't seem to bother her, nor did it bother me as she went about her cleaning work.

Vinh was about thirty years old and five feet tall. She had five children, two of which had died early. With long black hair, as most Vietnamese women had, she spoke only a few words of English, but seemed to understand much more. Rumor had it that her husband was an ARVN soldier, so we always felt she was trustworthy and let her have access to everything we kept in the hootch, even if we weren't there. She always wore silk clothing, usually with matching pants and top, sometimes blue, sometimes white or black, and other times wheat colored, along with black rubber flip flops.

With an innocent smile that revealed perfect pearl-white teeth, Vinh was as pretty as she was sweet. Like most hootch maids she was shy and avoided eye-to-eye contact which gave her a slight air of mystery. She was attractive, but untouchable, and we treated her with respect.

Vinh and the other hootch maids were paid 3,900 piastres (or maybe dong, if it had replaced the piastre by that time) a month, which was equal to U.S. $35. For this she was required to keep our hootch swept and our laundry clean, along with a number of other pilots' hootches. Al and I each tipped her $10 a month, which was a healthy amount by Vietnamese standards. No doubt she needed all the money she could get her hands on, because the cost of living in Can Tho City had gone up considerably due to the additional demand for goods and services brought about by the Americans' presence.

**C-Day**

Wednesday, 7 October was "C-Day" (C for conversion) when all the Military Payment Certificates (MPC, or Mickey Mouse Money as we called it) was turned in and replaced with new MPC with a different design and color. The MPC certificates were paper money in denominations of 5, 10, 25 and 50 cents, and 1, 5, 10 and 20 dollars. Each certificate measured two and one-eighth inches by four and one-fourth inches and felt like American paper currency to the touch. We used it for all purchases at the O Club and PX, and also to tip Vinh.

We were paid in MPC because Americans in the military service were not permitted to have or use American currency because dollars were more stable than Vietnamese piastres and more desirable to the Vietnamese. The American and South Vietnamese governments feared that lots of dollars in the Vietnamese marketplace would destabilize the native currency and create a black market for dollars that would affect the official exchange rates of the two currencies. This article from Wikipedia gives more details:

> C-days in Vietnam were always classified, never pre-announced. On C-day, soldiers would be restricted to base, preventing GIs from helping Vietnamese civilians—especially local bars, brothels, bar girls and other black market people—from converting old MPC to the newer version. Since Vietnamese were not allowed to convert the currency, they frequently lost savings by holding old, worthless MPC. People angry over their MPC loss would sometimes attack the nearest U.S. base the next night in retaliation.
>
> To illustrate the Vietnam War MPC cycle, in mid-1970, a GI could have a friend in the United States mail a $100 bill in standard U.S. currency, take it "downtown" and convert it to $180 MPC, then change the MPC to South Vietnamese piastres at double the legal rate. The soldier could then have a day shopping, bar hopping, or otherwise spending freely, paying in low-cost local currency, and finishing the day with a hefty profit.
>
> To continue the black market cycle, that $100 greenback would find its way to high-level Vietnamese government officials, especially the corrupt ones, who could travel out of country, where the U.S. currency could be deposited safely (Bangkok, Taipei, or Hong Kong). Rumors also suggested that this hard currency (US dollars), would find its way to North Vietnamese European exchange accounts.

As I recall there was only one C-Day during my tour, and it was not popular with anyone. Although it was not pleasant, there were no repercussions from the Vietnamese.

--- Ω ---

Less than a week later there was another early morning shakedown of the EM barracks which yielded a stash of marijuana in a refrigerator. The last time guys had been caught with it they were busted to Private (E-1, the lowest EM rank), fined $80 a month for three months and confined to hard labor for 2 months. One of the men busted that morning was scheduled for DEROS that day, but the poor sucker would be stuck an extra month for having dope. I guess there was more than one type of dope in that room.

One of the MPs who did the arrest told me they had busted a GI in downtown Can Tho the night before who had an unopened pack of Viceroy cigarettes that actually contained marijuana, not regular tobacco. It was amazing how the black market could duplicate the packaging of American cigarettes with marijuana inside and make it look brand new.

--- Ω ---

The same week as the shakedown occurred, one of the men on night shift in the maintenance platoon pulled me aside and told me a sad story. He had received a letter from his mother saying his wife was having an affair with another guy back home. His mom wanted him to divorce his wife, a German girl he had married while he was stationed in Europe, but he just couldn't bear the thought of losing her. She had admitted her mistake and wanted to continue the marriage, and he wanted to go home to mend the relationship. The Army refused to give him leave because too many GIs were using marriage problems as a reason to leave Vietnam and not return. So, he was stuck. I didn't know what to say when he told me the story, but I felt very bad for him.

--- Ω ---

A night or two later someone threw a lit incendiary grenade inside the right cowling of one of our M model gunships while it was parked on the flight line. The grenade, filled with thermite that burns at an extremely high temperature, melted a hole through the engine deck. We were all amazed that the Huey didn't blow up or burn to the ground. A brave passing soldier from a cavalry unit put the fire out.

--- Ω ---

It was always hot in Can Tho, and my clothing was usually soaked with perspiration. By October I was used to the heat and felt uncomfortable being in air conditioned places like the O Club. A shower was hard to find time for, but when I did it was a welcome relief. However, as soon as a shower was over the sweating started again, and by the time I got to putting on clean clothing I was drenched in sweat. Five minutes later it was as if I had taken no shower at all. Can Tho was a hard place to keep clean.

--- Ω ---

One evening about the middle of October, Denver Stiles was playing cards with Henry Solomon, Gordon Clayton and another pilot when all at once Stiles had a seizure. Clayton stuck his thumb into Stiles' mouth to prevent him from choking on his tongue, but Stiles bit down hard and Clayton couldn't get it out until they had carried Stiles to the dispensary and the doctor pried his mouth open.

Stiles eventually came out of the fit and was sent to a hospital in Saigon where an EEG revealed he was epileptic, thus ending his flying career. On top of all that difficulty, he received notice while in Saigon that his wife and child were ill and he needed to return to the States at once. This was a ruse by his wife, and the real story involved another case of infidelity and divorce. In the space of a couple of days Stiles had lost his family and his flying career. It must have been a terrible shock.

**Tigers Go Home**

I knew virtually nothing about Soc Trang, which was a strategically important, medium sized city in Soc Trang Province, which bordered the South China Sea southeast of Can Tho. The name of the city meant "silver depository" in the original Khmer language, because the Khmer king of old times stored his silver treasure there. At one time the city had also been called "Moon River."

I knew where Soc Trang was and had landed at the U.S.-controlled airfield there once or twice while on ash & trash and CA missions. In spite of my relative ignorance about the place, Soc Trang had significance to me as the home base of the 121st Assault Helicopter Company, the Soc Trang Tigers, the unit to which Tom Yost was assigned. It was also the home base of my classmate, 1LT Jack Bagley.

The 121st had a distinguished history as the first helicopter company in the Mekong Delta, relocating there from Da Nang in September 1962. In January 1963, the unit was presented with a Bengal Tiger cub which remained with the unit for only six months (imagine a wild tiger as a mascot!) then was relocated. The legacy of the temporary mascot was its name, and the unit was forever after known as the Soc Trang Tigers.

In October the 121st was to be deactivated and its duties turned over to the Vietnamese. But before that happened, all the company's personnel with significant time left on their one-year tours of duty were reassigned to other units throughout Vietnam. Tom was reassigned to the 13th Combat Aviation Battalion, the parent unit of the 191st, and so he came to Can Tho when the battalion headquarters was relocated there from Soc Trang. Although we had seen each other only a few times between May and October, we took up our friendship again as if no time had passed at all.

**Remembering Jack**

By Tuesday, 27 October, the 336th Assault Helicopter Company in Soc Trang also had turned over all its equipment and missions to the Vietnamese Air Force (VNAF). There was one mission left for American pilots to fly that day, the firefly mission near Soc Trang, and my classmate, Jack Bagley* volunteered to fly it. What happened that night was tragic and a surprise to all of us who knew Jack. The story that follows was written by Tom Wilkes*, a member of the 336th who was there that night:

> Jack was a respected pilot in our company and everyone liked him. As I remember he was the 1st Platoon Leader for the Warriors slicks. Towards the end of October, 336th was getting ready to turn

our company area over to the ARVN. We were under-strength as a lot of the company had already been transferred to other units. Super slick's AC, crew chief, and other gunner had already left 336. They took the Nighthawk set-up off the ship. That ship along with another, with two brand new crews, was assigned to just fly a very limited AO around the airfield. I had just gotten back from R&R and was flying combat assaults and they had me patrolling the company area that night. I think they wanted to make sure the ARVN didn't steal anything. Jack's ship was flying while the other ship was parked on the airfield. They were rotating 2 hr shifts.

All of a sudden an officer came running out of the operations hootch yelling to me that Jack had crashed and that he couldn't get the other ship on the radio to scramble them off. I said I would run out to the flight line to see what was going on. I ran out and alerted the crew but there were no pilots at the ship. The crew told me that the pilots had gone back to the company area. I ran back to the company area, located the pilots and the three of us ran back to the flight line. I helped them get the ship up and as the ship was lifting off someone grabbed me and pulled me aboard. One of the pilots had said, "Get Wilkes on the ship."

We flew to the area which wasn't far from the airfield and could see the ship burning. We landed in an LZ and I and two other crew members jumped off the ship. The ship then took off to drop flares. We were too far away from the crash to see it but knew which direction to go. The other two started running but I called them back. I came from the infantry and knew you just can't run through the jungle or we might all be dead. At that point we still had no idea of how the ship crashed or how many enemy were in the area. I guess because they were a lot newer than me they seemed ready to take direction from me. We then proceeded to patrol to the crash site. I can't tell you how much it has haunted me over the years about the decision I made to not run. Could we have saved them if we ran to the crash site? Logic tells me that I was doing what I was trained to do and that the ship had already been burning when we landed. Still the "what if" haunts me to this day.

When we got to the crash site we located two survivors, one badly hurt and the other banged up pretty good. We tried to get to Jack and Pysz but the fire was too intense. It was obvious that both were dead. We started first aid and I had them set up a small perimeter away from the burning ship. Another ship then landed closer to the crash site and six additional people assisted us with first aid and security. A medevac came in and extracted the wounded crew members. The ship that inserted the additional six then returned and we were all extracted.

As we were lifting off I noticed that there was no pilot in the left seat so I jumped into it. I was worried that if our pilot was shot there would be no one to fly the ship and we would also crash. I put the flight helmet on and briefed him as to the situation on the ground. We had no communication since we had initially inserted. When we returned to Soc Trang the others got off the helicopter and we got a new copilot. I stayed on the ship and we returned to the crash site. By that time ground troops had gotten to the crash site and had it secured.

We stayed on station to provide air cover and were then told that they had recovered the bodies. We landed and picked them up. I remember helping to load them aboard and that was a chilling sight as they were burned horribly. We then flew back in compete silence. I remember meeting with one of the crew members after he got out of the hospital along with the first ones to get to the crash site with me. He said that they were low leveling and turned out the light to keep the element of surprise and that was the last he remembered. We were all shocked, saddened and stunned by what had happened because we had all thought we were leaving Soc Trang without experiencing another death since we lost six crew members in February.

Thomas J. Wilkes

Apparently Jack had banked his Huey at low level with the light turned off and never saw the tree that killed him and his door gunner, SP4 Alex Pysz*. War, so arbitrary, so unforgiving, had taken my classmate and friend from Marshfield, Missouri.

**War Correspondent**

October in the Northeastern U.S.: frosty mornings, scarlet Maples, the aroma of burning leaves, raucous football games, and jack-o-lanterns. But in Vietnam there were no changing leaves or frosty mornings. It was always hot and humid, leading to more of the same, the monotony of predictable weather providing contrast to the unpredictability of war.

Jack's death was depressing, but late in October I was surprised and pleased to receive a small pumpkin and some Halloween candy from my wife. It was a real pumpkin and when the news of its arrival got around, lots of pilots in the company came by to see it.

Tom had seen the pumpkin too, and one afternoon he brought a SP4 named Mike Tschudi* to see it. Mike was a public relations specialist (reporter, or war correspondent) with the 13th Combat Aviation Battalion who was looking for stories to write for *Stars and Stripes* newspaper. Tom thought a "pumpkin from home" would make a good story, so he brought Mike in to interview me about it.

They came into the hootch and sat down, and after introductions were made Mike admired the pumpkin and asked if it would be alright if he wrote a story about it for the newspaper. I was flattered that anyone would consider a story about me for *Stars and Stripes*, so of course I agreed. We talked for a few minutes and Mike took notes for his article, and then hurried out. After he was gone, Tom filled me in on Mike's story.

Michael G. Tschudi was from Jensen Beach, Florida and had been drafted into the Army after dropping out of Catholic seminary. Following basic training and infantry AIT (Advanced Individual Training) he was sent to Vietnam and assigned to the U.S. 9th Infantry Division in the delta. After several months slogging through rice paddies with an M-16 rifle, he became the radio operator for his company commander and then had to slog through rice paddies with an M-16 and a PRC-25 radio (usually referred to as a prick 25) on his back.

Being a radio operator in an infantry unit had a couple of big drawbacks. To start with, the radio weighed nearly 25 pounds and that much weight, in addition to all the other gear that had to be hauled around (nearly 70 pounds worth), would wear a man down quickly. Worse than the extra weight, however, was the need to stay close to the company commander so he could use the radio whenever it was needed. This could be uncomfortably restrictive, or even dangerous during a fire-fight, because the VC would look for the radio operator and the company commander who was always nearby, and try to kill them first. Hauling a PRC-25 was like wearing a bulls eye.

Mike had been reassigned from field duty with the 9th Division to the 13th CAB when the 9th's duties were transferred to a

Vietnamese infantry unit. All the 9th's personnel with significant time left on their tours were assigned to other units, and Mike was assigned to the 13th CAB to be a public relations specialist because he had writing skills and some college education.

The oldest of 17 siblings, Mike was slim and sandy-haired with a neatly groomed mustache and soft brown eyes. He was a chain smoker who preferred wine over alcohol and beer, and to my knowledge never used marijuana or drugs. He was engaged to be married to a young woman back in the States and was excited to be going to meet her on R&R just a few weeks after we met. And in spite of the prohibition against it, he kept a journal in which he recorded his daily combat experiences, including the most harrowing ones with the 9th Division. The journal also contained some of his original poetry which he occasionally read to Tom and me. (See one of Mike's original poems in the "Holly Holy" story in December.)

As I got to know him better, it became clear that Mike did not plod through life with an obvious logic. Instead, he felt his way through it using a well-developed sixth sense with which he could see things in others that they could not see in themselves. He was interested in what made other people tick, but I was interested in what they did and whether their judgment could be relied on when the going got tough. Such contrasting approaches could make for unsettling encounters, which we had from time to time, but by and large we got along exceptionally well and enjoyed each other's company.

Shortly after we met, Mike went on R&R to meet his fiancée. He was giddy with excitement before he left, but after coming back he told us she had returned his engagement ring and broken off their relationship. He hadn't seen it coming and was crushed. I felt terrible for him and didn't know what to say. Such a development seemed worse than a "Dear John" letter, and it appeared to strengthen the tendency toward melancholy that always seemed to underlie Mike's perspective at that stage of his life. (A "Dear John" letter was a letter written to a husband or boyfriend by a wife or significant other, ending their relationship, usually because the writer had found another lover.)

There was no way of knowing that Mike and I would be friends for over 30 years following our meeting in October 1970, or that he would pass on in October 2002 at age 53 while serving as the Rector for a Catholic Cursillo spiritual retreat in Florida. He was a fine man and I miss him.

**For A Few Dollars More**

Five years before I arrived in Vietnam a "spaghetti western" with the above name was made starring Clint Eastwood as a bounty hunter who was tracking down a fugitive in the story. ("Spaghetti western" is the nickname for cheap movies made in Italy about the American West.) How the Third Platoon of the 191st came to be called the Bounty Hunters is unknown to me. I suppose they may have believed the title of the movie matched their desired image as tough killers who worked for pay, and that they were good at what they did in fearless fashion.

In the air, we slick pilots didn't care what the Bounty Hunters thought or did as long as they were where they should be when we needed cover going into an LZ. And that, they were good at: prepping LZs with rocket fire, guarding the flight on every insertion and extraction, and staying at the LZ to pound the enemy into submission after the flight had left it.

In spite of the rudimentary crudeness of their weapons, which were limited to machine guns and rockets that were bore-sighted (i.e. aligned with the axis of the Huey's fuselage), the Bounty Hunters were good at what they did. Their sighting mechanism was an X marked on the inside of the windshield with a grease pencil, and when the X lined up with a target, they squeezed off a rocket. Being bore-sighted, the rocket would go in a straight line in whatever direction the fuselage was pointed. Or at least that's what I thought until one of the gun pilots showed me how they could bend the flight path of a rocket in mid-air, an amazing feat to me.

Each Bounty Hunter Huey had an emblem painted on the nose cone containing the words "For A Few Dollars More," and all the Bounty Hunter crew members, including the enlisted men, wore red bandanas to distinguish themselves from the rest of the air crews. Perhaps the ultimate distinguishing feature, at least among the pilots, was the Bounty Hunter Lounge next to the pilots' hootches, almost directly across the walk from our hootch. Only Bounty Hunter pilots were allowed in the lounge, unless invited in by a Bounty Hunter.

Most of the time the pilots and crew members from the other three platoons didn't think about the distinction the Bounty Hunters tried to maintain. But once in awhile something would happen to aggravate the relationship between them and the rest of the company members, and hard feelings and resentment would arise. So in order to keep the peace, the Boomerang and Green Delta pilots decided to construct their own lounge which everyone was welcome into without question.

An ideal location between two Boomerang officers' hootches was identified and everyone started collecting the materials for the lounge: wood, nails, steel roofing, a refrigerator, furniture, etc., and within days the new Boomer Lounge was open for business. It was about 20 feet square, fully enclosed with a door on the front and a bar off to one side. A bamboo couch and several chairs were found and placed around the outside walls and a couple of bar stools were added. The inside walls were finished in pine paneling, a ceiling was installed and lights were put up in the corners. Where all that stuff came from I didn't know, but a determined soldier can scrounge up just about anything.

With no written guidelines or agreements, a general fund was established from open contributions based solely on an informal understanding of how the money would be spent. General fund contributions were used to purchase an initial stock of beer, wine, liquor and soft drinks, and from then on proceeds from the sale of drinks replenished the general fund. It was quite amazing how smoothly it all played out; I don't recall a single quarrel or disagreement over the operation of the lounge the entire rest of my tour.

"For a few dollars more," and with focused determination, we had a lounge of our own with a genuine Australian boomerang mounted on the wall behind the bar. Within the confines of our own hootches, we now had a place where pilots could blow off steam, tell war stories, or listen to their favorite music on the stereo someone contributed for the good of the cause. The lounge would turn out to be a busy place!

**After Midnight**

One afternoon about 4 PM, I decided to go to the Boomer Lounge for a beer. For a few minutes I was there alone with my thoughts, and then Mitch Gregg dropped in. We talked for a few minutes and soon other pilots began to straggle in one-by-one as their daily missions ended.

Al came in a few minutes later and ordered a beer, and we stood next to each other at the bar making small talk. An hour or two went by and we began to feel quite mellow as one beer followed another. Then all at once Al stepped back and looked me up and down as if he was inspecting me. He reached over and grabbed the edge of a patch on my uniform sleeve that had come loose, and with a cynical smile ripped it completely off my shirt. "There," he said with satisfaction, handing the patch to me, "now get that sewn on right."

Knowing this was done in jest I was not angry, just a little miffed that I would have to take the time to have it fixed. Al obviously had been looking for something to do, and I was the closest one to do something to, so I was the target of this practical joke. And it wasn't much of an issue because I would just have the local Vietnamese tailor sew it back on the next day. Gregg thought it was hilarious as did a couple of the other pilots, but that's as far as it went.

| *After Midnight* - Eric Clapton |
|---|

We continued to drink throughout the evening, with Al occasionally looking at my uniform for more targets of opportunity. Nothing happened until slightly after midnight when most of the pilots in the lounge were well beyond mellow.

Al began picking at my uniform and then grabbed the edge of my name tag, tearing it off and saying in a slurred mumble that I was really out of compliance with uniform regulations now, and as an officer and gentleman I should be ashamed of myself.

It was late, and the other pilots in the lounge were soused enough that Al's initiative sparked a mob reaction. In an instant Gregg had grabbed the U.S. Army strip over my left pocket and ripped it off. Then someone else behind me grabbed my back pants pocket flap and ripped it off, and it was frenzy. Within the space of half a minute the mob had done its work. I found myself standing at the bar with nothing on except what couldn't be torn off: my boots and socks, my belt with a couple of strips of olive drab cloth hanging down from it, my boxer shorts with one cheek torn out of the backside and my dog tags. The place was in an uproar!

All throughout the evening I had lamented my desire to leave Vietnam and go home, and as more cheer was consumed this expression became more emphatic. The uniform frenzy had brought these feelings to a head, so in a fit of momentary insanity I shouted to the crowd that I was going to 164$^{th}$ Aviation Group Headquarters to demand I be sent home at once, and strode out of the lounge to a chorus of hoots and howls.

I ran down the sidewalk to the group headquarters building a few blocks away, pausing on the front sidewalk to collect myself and wipe off the sweat pouring down my face and chest as a result of too much to drink, and too much heat and humidity. Marching up the walk to the double glass doors I could see the officer of the day, a young second lieutenant in a khaki uniform, seated at a desk about 15 feet inside the entrance looking like a picture-perfect Army officer.

I pushed the door aside and marched up to the desk as he raised his eyes from a book he had been reading. Sweat was streaming in my eyes and dripping off my dog tags, making a puddle on the tile floor around my boots as I stood at attention. When he caught full site of me his eyes widened, and with a look of confusion on his face he stammered, "Is … is there some …" (looking me up and down) "something I can do for you?" I replied, "Yes, I would like orders to be cut right now sending me back to the States. I've had enough of this war crap!" Nearly dumbfounded he said, "What was that?" and I repeated what I had said.

The expression of shock on his face quickly changed to suspicion, and I noticed his right hand slowly reaching for the telephone on the desk. He asked my name and I slurred something I can't remember. His expression changed to a scowl as his hand moved faster toward the phone. My eyes followed his hand. As he picked up the receiver and started to speak, I turned around and took off out the front door with the lieutenant yelling, "Stop! Stop!" after me. But I was not about to stop and be arrested.

Although the airfield was nearly deserted at that time of day, I didn't want anyone to be able to follow me to the Boomer Lounge, so I took a winding route back and found the impromptu party still in full swing. When I explained where I had been and what had happened, the other pilots howled with delight! I had dared to do the outrageous and came back to tell about it. Boomerangs always some back!

The party went on and on until the wee hours, with the pilots gradually trickling out, headed for some much needed rest before morning missions started. I don't remember when I hit the sack, but I do remember that the next day I had a terrible hangover ... and was short one flight suit!

**Topping Check**

One of the routine checks for a Huey that was required at every PE was an N1 topping check. The N1 was the primary turbine inside the jet engine, and the topping check was a diagnostic procedure to measure its efficiency. In essence, it was a procedure to check the power and condition of the engine.

A few nights after my clothes were torn off in the Boomer Lounge, I found myself on a test flight in a Huey, surrounded by pitch dark and climbing past 10,000 feet in the middle of a topping check. There was no moon, but a few faint stars could be seen between the scattered clouds. The red light from the rotating beacon on the roof throbbed above me. Each time it went around, it bathed the cockpit in a red glow as light reflected back into my eyes from the clouds surrounding the Huey. The air was cold and clammy, and soaked in a mixture of humidity and nervous sweat I just wanted this test flight to be over!

My stomach was in knots, my legs and feet rigid against the pedals. My hands gripped the stick and the collective as tight as possible in doubtful hope that I could hold the aircraft and me together by physical force and willpower. My eyes danced between the instruments on the dashboard and the airspace outside. The threat of vertigo came and went in waves.

"Robertson, is your equipment all set?" I asked the technician in the cargo compartment behind me. "Yes, sir," he answered as the powerful new Huey continued its climb at 1,500 feet a minute. "This new Huey with its brand new engine is relentless," I thought to myself as we passed 11,000 feet. Plunging into another cloud the rotating beacon's red rays again filled the cockpit. I wanted to turn the damned thing off so the vertigo it caused would stop, but the possibility of not being seen by a much larger plane at this altitude, at night, overrode the temptation. I'd been given no official clearance to be at such high altitude where the super zoomies (i.e. the big jets) roamed, and while vertigo was bad a mid-air collision would be worse.

I continued climbing at maximum power, which was the required procedure for an N1 topping check, and two minutes later passed 14,000 feet where at last the Huey broke out of the scattered clouds and into clear air. With no clouds to reflect the light of the rotating beacon, the vertigo vanished instantly. And with that distraction gone, I felt the air getting uncomfortably colder. Breathing became noticeably more difficult as the oxygen content in the air thinned out, and I could see my breath in the frigid air. This high altitude was where we were not supposed to be without extra oxygen, and the fear of blacking out seeped into thought, displacing all the other fears and distractions. "When is this bitch going to top out?" I wondered.

I glanced at the altimeter and then the tachometer, waiting for the engine RPM to go down enough for Robertson to get his required instrument readings. I glared at the tachometer hoping my stare would force the RPM needle to go down. But then I saw the altimeter pass 14,300 feet and my anxiety jumped another notch. I had never flown a Huey this high and there was no indication as to how much higher I would have to go, but I badly wanted the climbing to end.

My eyes moved quickly and repeatedly in a tight pattern from the altimeter to the attitude indicator, then to the tachometer. With heightened expectation my eyes flashed back and forth as my grip on the controls tightened even more. When the altimeter hit 14,500 feet I thought I detected a slight drop in the tachometer. I looked again ... and yes! I could see a drop of 25 RPM; then 50!

The altimeter said I was now passing 14,800, but the tachometer seemed stuck. 14,900 came and went. Passing 15,000 feet I hit an unexpected cloud and the vertigo from the rotating beacon came roaring back, the dizziness making me forget the frigid air. It was a mental battle to resist the temptation to lower the collective and head for the airfield. "If I don't finish this now," I thought, "I'll just have to do it over," and I figured one topping check in this chopper was enough! I definitely didn't want to abort at this point, no matter what my senses were telling me.

The tachometer was now down 75 RPM and I began to silently demand that this bird give up the last 25 I needed to end the flight, either RIGHT NOW, or IMMEDIATELY, whichever came sooner!! It was not rational, but neither was doing an N1 Topping Check in a brand new Huey at night at this altitude. For that matter, almost nothing in Vietnam seemed rational!

I glanced again at the tachometer looking for the magic number, and there it was! 6,500 RPM! "Robertson," I said as calmly as I could, "that's it, take your readings." "Yes, sir," he replied. But before he was done we had climbed another couple of hundred feet in a 20 second period that seemed like an hour. "That's it, sir, we got it," he said, and at that moment those words were like a sweet release from prison! The anxiety bubble burst as I lowered the collective and headed for Can Tho Airfield. This routine test flight and the stress that came with it were finished!

My flight suit was soaked with sweat, yet I was freezing from what seemed like arctic temperatures up at high altitude. I didn't ask Robertson how he felt, but I could tell by the tone in his voice he was relieved too. As I banked toward home field, I could see the bright lights of Can Tho City beneath a few scattered clouds, and I wondered which ones of them we had flown through. The relief I felt at being done with this test flight was off the chart, unmeasurable, and comparable only to what I felt when leaving a hot LZ in one piece. I knew the flight had been into uncharted territory, beyond the limits of anything I ever expected to experience in a Huey, and that knowledge gave me satisfaction and confidence. I could now understand how Chuck Yeager felt when he broke the sound barrier. We were soul brothers.

Before I became a Wingnut, what I knew about test flying came from exciting stories about great test pilots like Charles Lindbergh and Chuck Yeager, who had flown to the edge of space. Some of the Apollo moon crews of the 1960s consisted of test pilots too, so I fancied myself to be in a special class as a test pilot, especially after extended maintenance tests like the one described above. I figured that being a test pilot, even a maintenance test pilot, placed me in a fearless fraternity of men who risked their lives to break barriers that were both physical and mental. In a world of such men, of men who faced their fears alone and didn't back down, test pilot was a desirable title with good duties, even if it did scare the hell out of me!

--- Ω ---

On Wednesday, 28 October, about 1:00 PM, as I was sitting in my hootch, Jack Wyatt and Dave Mahala dropped in to talk. Jack had lived with us and Tom Yost in Savannah, and Dave too was in class 69-50. He wound up being assigned to the same unit Jack was in at Vinh Long, D Troop of the 7$^{th}$ Cavalry. They dropped in because their unit's operations were curtailed for a day for maintenance on their aircraft, and that gave them the opportunity to come to Can Tho for something different to do.

Jack and Dave both looked great and said they felt good too, but were tired of the war and wanted to get back to the States. Dave's wife, Jane, was expecting a baby on Nov. 15$^{th}$ and he was looking forward to getting leave to go home for the birth. He said Terry McLaughton, another classmate, was stationed in Phu Loc, about 20 miles north of Da Nang in I Corps, and was doing the same thing I was: working in maintenance. Classmate Ted O'Bean was flying Cobras for the 101$^{st}$ Airmobile Division in I Corps and was sending medals home by the boatload. And classmate Don Basil, who I mentioned seeing at Sanford Field, had been grounded recently after he crashed attempting a hammerhead stall in an OH-58 (a Bell Jet Ranger) trying to impress a girl in the backseat. He was now flying a desk instead of a helicopter.

# Chapter 7 - NOVEMBER 1970

Only 180 days to go.

A full moon would rise on Friday, November 13th.

***Fight the Team* (The Ohio State Fight Song)**

Ohio State was #2 in the football rankings going into November and had won all their previous games that season. I tried to follow the team's progress by whatever means were available and could rely on *Stars and Stripes* to at least report the game scores. My wife was ever faithful in sending newspaper clippings of the games, which was great except that they came a week after the games ended. But on Sunday morning, November 1st, AFVN radio carried the Ohio State-Northwestern game, live from Ohio Stadium.

Northwestern was ranked #20 at that point and had a good team, so Tom and I were eager to listen to the broadcast. I had gotten off duty at 11:00 PM because of a red alert that required "lights out" across the airfield. With no lights, no maintenance could be done, so I went back to the hootch to read by myself while everyone else was suffering in the bunker. At 2:15 AM the game came on and I listened to it at the hootch until halftime, then went to the battalion operations center (BOC) where Tom was on duty.

The BOC was actually a small dark building with no windows, half buried under sand bags to protect it from mortar rounds. It was a short walk from my hootch, and Tom and a couple of EMs were the only ones there when I arrived. He had the game on the radio already, and when the second half started we were giddy with anticipation. Listening to the marching band and the roar of the crowd in Ohio Stadium made us grievously homesick, but it was a great way to start the month.

Eventually the Buckeyes prevailed against the Wildcats 24-10 and we celebrated the victory the following night with popcorn and beer. OSU went on to win the rest of their games in November and were selected to represent the Big 10 in the Rose Bowl on January 1$^{st}$.

**Nemesis**

November had arrived and with it the weather started to change as the monsoon season waned. But it was still hot and steamy ... it was always hot and steamy ... and the only respite was to jump into a Huey and climb to 1,500 feet to reach cooler air, or go to the officers club where the air conditioning was so cold it would chill the fuzz off a peach.

The workload had been light for a couple of days, with only a few test flights and a little record keeping needed. So, one day toward mid-afternoon I went to the O Club to cool off and get a beer. The club was a haven for officers from all over the airfield, and especially for pilots who were off duty. In addition to being air conditioned, one could always count on a cold ham sandwich and boiled eggs being available on a moment's notice, seven days a week. And of course the beer, wine and alcohol were always in high demand.

The one-story club building was modest, being roughly 60' long and about the same in width. The exterior was wood with the upper part painted white and the lower part redwood. The inside was wood paneling, and there was a long bar along the wall to the left of the entrance. There were about twelve stools at the bar, and two dozen small tables, each one surrounded by four wooden chairs with black vinyl seats and backs. The lighting was usually subdued, even during the daylight hours, except for a cement patio at the rear that was always awash in sunlight. Against the side wall opposite the entrance was a stage where floor shows performed, but it was dark and quiet now.

As I walked in from the bright daylight the darkness inside left me momentarily blind, and I had to feel my way to the bar. It was quiet and even in momentary blindness I could tell there was almost no one in the club. As my eyes adjusted to the low light, I could see several Vietnamese waitresses talking with the Vietnamese bargirl.

I sat down at the bar, waited for my eyes to adjust to the light, and ordered a beer. As my eyes got used to the dark, I could see Tom Yost seated at a table with another officer who was unrecognizable. When the beer arrived I made my way to Tom's table, and as I neared it Tom said, "Dick, do you remember Major Jarman from ROTC at Ohio State?" I was speechless, completely surprised that the officer I thought was the epitome of a useless lifer had shown up at Can Tho Airfield, and was talking with Tom.

At Jarman's invitation I sat down uncomfortably across from the man I had despised as an undergrad. Our ensuing conversation, however, showed he was not quite the belligerent jerk I knew at college, maybe because we were all officers at that point and there was no need for the faculty-student games he used to play. Beyond that was the fact that I was a Huey pilot, taking risks everyday for "the cause," and he was a desk jockey somewhere. At one point in our conversation he said, "Yeah, Lt. Barth, I didn't think you were going to make it." And I responded that I didn't think I was going to make it through ROTC either!

I drained my beer and, after a few more minutes of awkward conversation, excused myself and left. What it was I didn't like about Jarman amounted to two things: a lack of compassion for people below his rank and an attitude of snobbish superiority. I hoped our paths would never again cross, and they haven't. In spite of his personality, I suppose his mother loved him.

**Bad Moon Risin'**

*"Don't go 'round tonight, it's bound to take your life,*
*there's a bad moon on the rise."*

John Fogerty

| *Bad Moon Risin'* - Creedence Clearwater Revival |
|---|

The required minimum tour of duty was 12 months, but some pilots were just not ready to return to the states when their year was up. Jack Orlo was one of those pilots. He had arrived in August 1969, and 14 months later he was still flying for the 191st. Only now he was no longer flying the lightship in the first platoon, he was a Bounty Hunter flying a gunship.

On Monday morning, 2 November, he was flying with Warrant Officer Tom Lauper who had transferred to the 191st from the 175th Assault Helicopter Company, the "Outlaws," in Vinh Long. Tom was 21 years old and married, with a home in Orlando, FL. He graduated from flight school in class 69-45, a couple of classes ahead of me and a few more behind Orlo.

That morning, Orlo and Lauper were flying a Huey C model gunship with tail number 66-15219 that had been purchased by the Army in September 1967. It was not an especially old Huey, having only 2,635 flying hours on the frame. Flying with Orlo and Lauper were the crew chief, SP5 Ken Nichol from Kansas and the door gunner, SP5 Ockie Mahoney from Arizona.

At 6:50 that morning, they took off from the Rach Gia short strip headed for Rach Soi Airfield (where Mahan and I were ambushed) to refuel. They were accompanied by another Bounty Hunter gunship and crew, also on their way to refuel at Rach Soi. Reaching 500 feet, the transmission that drove the main rotor seized, causing the main rotor head and rotor blades to twist off the shaft that connected them to the body of the Huey. With no rotor blades the Huey nosed over and dove 500 feet almost straight down, bursting into flames when it hit the ground in a rice paddy. All four crew members perished. If Orlo had returned to the States when his tour was up, he would have been safe at home on 2 November.

Losing lives was always regrettable, but losing lives in combat as a result of a mechanical failure was an utter waste, and when the details of what happened came out there was more to the story than met the eye. The pilots of the other Bounty Hunter gunship flying with Orlo and Lauper reported that Orlo radioed them before taking off from Rach Gia saying he had a transmission warning light that was on (indicating a problem with the transmission), but the transmission temperature gauge was normal. He decided to ignore the warning light and turned it off, a no-no from flight school training and a deadly decision.

The loss of those four men was pitiful, but a loss due to bad judgment was even worse. Our flight instructors had drilled into us the importance of paying heed to what the instruments and warning lights were trying to tell us, and to never turn a warning light off. Why Orlo did that is a mystery.

I had flown the Nighthawk mission with Jack many times when I was in the Boomerang Platoon. On 2 November I was grateful that his path and mine had diverged and I was not his copilot that morning.

When the wreck of that Huey was returned to CTA later that day, a hairline crack was found in the main rotor mast. I looked at the crack and saw rust around it indicating the crack had been there for some time. Some speculated that somehow the mast had broken in flight causing the accident, but this hypothesis didn't seem to line up with the story about the transmission warning light.

The question of what actually happened was never answered. Maybe a bad moon had risen.

**Dinky Dau (*dinky dow*)**

Every Army company has a 1st sergeant, usually known to the rest of the company personnel as "Top," which is short for Top Sergeant. Our Top was a quiet man who I barely knew because we never had any reason to do business together. From my point of view, Top was at the company headquarters doing his top sergeant duties, whatever they were, and that was that. It wasn't something I thought about.

But that changed in November when I walked by the enlisted men's barracks and saw Top mopping the floor in the latrine and singing quietly to himself. Mopping the floor was a job reserved for low ranking enlisted men, or hired Vietnamese, but certainly not for a 1st sergeant. I asked around to see if anyone knew what had happened that he should be mopping the latrine, but no one seemed to know. The implication was that he had lost his marbles, or was dinky dau as the Vietnamese would say. A short time later a new 1st sergeant appeared in the company to take the old Top's place, but the old Top didn't leave, he just continued to mop the latrine.

I never knew anything else about the situation. It was mysterious, a secret that no one would talk about, strange even by Vietnam standards.

**Unspeakable**

There were stories about it that circulated throughout the press, but no one in the Army talked about it. It was too threatening, and discussing it just accentuated a vulnerability that brought great discomfort. But fragging was a real problem, and one that was safe to consider only when it was in another unit, or far enough away not to be a threat.

Fragging was the act by an American of killing another American using a fragmentation grenade, and every person in authority in Vietnam had an unspoken dread of being fragged by some angry and/or hopped-up GI. This included commissioned and warrant officers as well as higher ranking Non-Commissioned Officers (NCOs), and most of the stories in the public press reported fragging incidents as taking place in infantry units out in the jungle. The military police reported there were 333 cases of fragging in Vietnam in 1971 representing an epidemic of despair and disillusionment in the ranks.

In September a new Sergeant First Class named Jake Ozzimo had been assigned to the Maintenance Platoon as a technical inspector. Jake was of Italian descent and a real go-getter, a tough and demanding NCO who made the crew chiefs and door gunners live up to a higher standard of personal and professional conduct. He didn't think twice about berating any enlisted man who was out of line, yet he was respectful to all, especially to those who out ranked him.

Ozzimo, a 34-year-old whose home town was Columbus, GA, was thorough and methodical in his approach to his duties. So, every night after our Hueys had returned to the airfield, he would check their log books to be sure they were ready to fly the next day. His pattern was to open the small jump door behind the copilot's seat and reach in to get the logbook which was usually near the door. On 4 November, when he opened the jump door of one of the Boomerang Hueys, there was a terrible explosion from a fragmentation grenade that had been set as a booby trap. He was killed instantly.

For many days afterward, investigators from the Army's Criminal Investigation Division (CID) prowled the airfield interviewing members of the company and examining the damaged Huey for clues. To my knowledge the murderer was never found, but there were lasting effects as the officers and NCOs in the company became wary of the enlisted men, and the relatively little trust that existed among the Huey crews reached new lows.

We all understood that this act of barbarism against an NCO was an overt attack on the respect for authority that the Army ran on. This essential element was critical. Without it, the officers and NCOs would be unable to direct others in combat conditions, and that would compromise everyone's safety as well as the successful completion of our missions. This was an assault on the bedrock of the working relationship between those in authority and those under them, and that bedrock had been permanently damaged.

Ozzimo's murder was tragic and no one had the courage or could find the words to talk about it. It was too horrible to remember and too threatening for conversation, so it was ignored, buried, forgotten.

--- Ω ---

Vung Tau (*vung tah-oo*) is a city on the South China Sea coast about 100 miles north-northeast of Can Tho. Vung Tau (which means "anchorage" in Vietnamese) was a well-developed port city where U.S. Navy ships anchored in the harbor and provided sophisticated maintenance support, like rebuilding jet engines for Hueys. But it was best known as an in-country R&R center for U.S. troops who enjoyed its gorgeous beaches.

The 191st was said to have had a villa somewhere in Vung Tau that was used as a private R&R center, but no one I knew seemed to know exactly where it had been. None of the existing pilots had seen it, and the closest I ever got to it were the rumors that circulated in the company describing it as the next best thing to a resort.

Decades later I located a “map” to the villa on the 191st website, which is displayed there. It shows the location to be on Cap St. Jacques, the French name given to the rocky outcropping at the end of the Vung Tau peninsula.

**CMMI**

CMMI, an abbreviation for Command Maintenance Management Inspection, were four letters that scared the hell out of commanding officers. A CMMI was an annual assessment of the effectiveness of the maintenance program in an Army unit, and failing one could result in an officer being relieved of his command.

A CMMI was coming up soon and MAJ Simon was in a tizzy for weeks leading up to it, worried sick the company would fail and his Army career would go down the toilet. Unable to bottle up his anxiety, he spent many days roaming around the hangar, worried stiff and trying to pass his fear on to the rest of us. But we wouldn’t bite. All we wanted was to get the CMMI behind us so Simple Simon would go back to his office and leave us alone to concentrate on the primary task of keeping the Boomerang choppers in the air. Referring to the upcoming inspection, one afternoon Simon said to me, “Well, LT Barth, have you achieved perfection yet?” to which I replied that no one I knew had achieved perfection and it was unlikely I would either.

The CMMI was on Veteran’s Day, Wednesday, 11 November. Early that morning, a small cadre of inspectors dressed in starched, pressed fatigues showed up at the company headquarters, clipboards and pencils in hand. They were led to the hangar where they immediately began thumbing through paper records and poking into corners trying to find deficiencies to report. When the day was over, much to Simon's immense relief, we passed the inspection, but not by much. A passing grade was 70; we scored 70.45.

From the stories I've written about him, one may get the impression that I thought Blake Simon was a mean bastard, sprung from the same mold as Jarman and Sechrist. But that's not the case. The truth is he was incompetent as a leader of men, and had been placed in a role for which he was not cut out. He might have served better in a staff position where administrative skills were more important than being able to finesse men in battle (a traditional military leadership skill).

Simon's assignment to command the 191st was an example of the Peter Principle at work, a theory of management summarized by the statement that "employees tend to rise to their level of incompetence." While the rank of major gave him the formal authority needed to be a commander, his personality and mannerisms did not demand the respect that was the foundation of an equally important informal authority. To say it another way, he lacked the warmth, confidence and charisma that inspire men to follow. He had risen to his own level of incompetence and I just wanted to be away from the guy. Today, I see him as a victim of a system that failed to adequately take into account his strengths and weaknesses and assign him to appropriate duties. For anyone who is or has been caught in this dilemma, I have sympathy.

There is a P.S. to this story. The next day was the 162nd's CMMI, and during the inspection a cache of extra parts that the 191st had been hoarding when we occupied the 162nd's hangar was found. This brought immediate reaction. The 13th Battalion Commander was notified about the unauthorized parts, and he decided that "heads must roll." As a result, Simon's world quickly went to hell. The battalion commander rushed into the situation and relieved the technical supply officer on the spot, and followed that up by relieving Simon of his command of the 191st.

But then he changed his mind and didn't relieve him. Then he did, and didn't once more.

It was a zoo. When the dust finally settled the tech supply officer was made the scapegoat and Simon was off the hook. As always, the feces had rolled down hill, and in this case the tech supply officer caught it.

**Diamond of the 191st**

He was a tall, lanky Texan with a big attitude and an equally big grin to go with it when he arrived at the 191st to fill the executive officer's position vacated by Ted Everest a few months back. But when Don Jefferson DEROSed in November, CPT Jim Diamond became the new Wingnut 6. Fresh from helicopter maintenance school, Diamond Jim, as we called him, was brimming with confidence and an aggressive attitude that could put 99% of the Army officer corps to shame.

Slightly stoop-shouldered with thick lips and thinning brown hair, Diamond was a caricature of the modern day Texas cowboy, or at least what I thought one should be like. He was on his second, or maybe third tour of Vietnam, and had come through OCS from a number of years as an enlisted man and NCO.

To me, this new Wingnut 6 was a great deal more than a breath of fresh air, he was a whirlwind of it, and I learned very quickly that everyone knew exactly where they stood with him, all the time. He was open and straightforward, able to not only discern truth from BS, but to be outspoken in the declaration of it. His communication was artful in its simplicity, pointed and unabashed, caustic when needed. His directness and resistance to political correctness were appealing to me because they reminded me of my family.

Shortly after he took over the Wingnuts, Diamond cleared the deck and spelled out what he thought our working relationship should be like. He said two things that made a big impression: first, he was counting on me to teach him about the platoon and the people in it, so he would know who to trust and depend on; and second, he expected our relationship to be one of equals, and he would call me by my first name if I would reciprocate, but only when no enlisted personnel were present. If they were, we needed to maintain military formality by addressing each other by rank and last name to maintain respect, discipline and order.

This short discussion is probably the best start to a superior-subordinate relationship I've ever had. From that few minutes of discussion, I believed I knew where he was coming from, who he was, and what to expect from him in the future. Our relationship promised to be the complete opposite of many I had experienced so far in the likes of Jarman, Sechrist, Simon, etc. and I'm happy to say I was never disappointed in that regard.

Most of the time, Diamond Jim was affable and even-tempered. He handled problems easily and in a relaxed manner that reflected a man who was competent, confident and comfortable with himself. He was a professional who seemed immune to most of the petty drama of daily life. However, there was one issue that always riled him to the point of physical confrontation, and that was any attack on his integrity.

For example, one evening after I had crawled into bed, maybe a month after Diamond arrived, I heard some loud, angry talking outside the hootch, which was followed by the sound of people in a physical struggle. There was some cursing and name calling, and then the sound of bodies hitting the ground, which eventually forced me out of bed to investigate. Through the screen I could see two men fighting, then opening the door and shining my flashlight on the pair, I recognized Diamond and one of the other platoon leaders locked in a wrestling match. I yelled at the two of them to break it up and go to bed, and that's exactly what they did. But as he left I could see blood on Diamond's lip.

The next day I asked him what had provoked the fight and he said the "bastard" had accused him of not doing his job, and of course that was his tender spot. They had both been drinking at the O Club and the alcohol stoked the argument, which finally erupted into a fight. He was not in the least apologetic, but I got the feeling he was unhappy with the way things had turned out. No doubt he would have cleaned the other captain's clock if he'd had enough time, but I thought it was inappropriate behavior for both officers. I felt that with an enemy population outside the perimeter wire to contend with, we didn't need to make enemies inside the wire amongst ourselves, and I told him that. He grinned and showed no sign of resenting my comment.

Diamond Jim loved to be in the air. He was always eager to jump into a Huey and get away from the menial demands of ground duty. He was a warrior, and he wanted to be out with the other warriors doing his warrior thing. When each flying day was over and we were on final approach to the Can Tho runway, he would key the radio microphone and announce to the world that Wingnut 6, the proud bird with the OD (olive drab) tail was on final approach for a landing at Can Tho Airfield (this was a parody of the Continental Airlines commercial of that day which referred to the Continental jet as "the proud bird with the golden tail"). I think it was his way of saying, "Here I am world, Captain Jim Diamond is back," because each time he said it he would look over at me and grin from ear to ear, a happy cowboy riding his horse home from the range.

--- Ω ---

By 19 November the Wingnut Platoon was full of officers who were either just arriving (Jim Diamond, Robert McCoy) or preparing to leave (Don Jefferson, Will Moss and Roger Murdock). McCoy had been assigned as the new assistant platoon leader and was working nights with me, which let me off the hook for running the night shifts. This was something I appreciated because it made my life easier, and with McCoy on board I could focus almost solely on test flights.

About that time, the decision was made to start showing movies in the hangar for the night maintenance crew. Most of the night mechanics worked 12-15 hours a day. They had little time for entertainment, so they welcomed some time to relax with a movie. The first one that was shown was a comedy western called *The Ballad of Cable Hogue* which contained a few scenes of Stella Stevens dressed in next to nothing. It was a big hit with the guys and the decision was made to show movies in the hangar every other night for awhile ... only for morale purposes, of course.

--- Ω ---

Very few pilots had the opportunity to fly with battalion commanders and see firsthand how they handled their command duties in the air. Around 20 November, Tom was picked to be one of those pilots, and soon he was flying with the 13th CAB commander whose call sign was Guardian 6. This freed Tom from the boring duties in the TOC and got him out into the AO nearly every day. His new call sign was Guardian 6 Alpha, and while I got to see him less after that change, I was glad he was happy being back in the air.

--- Ω ---

From my letter dated 23 November 1970:

> ***Last night I was down on the flight line working on a ship and the sky was as clear as a bell. So I lay down on top of one of the revetments and just stared at the stars-- so cold and beautiful. With so much beauty in the world, why do people have so much trouble getting along? What's wrong with man? Seems it has to get better-- don't see how it can get worse. When a man is kept from fighting, from loving, from satisfying his desires, what do you have left? A lot of thought provoking questions I guess.***

**Unexpected Casualty**

Thursday, 26 November, was Thanksgiving Day back in the world, and I imagined Americans getting together to share both their bounty and their gratitude on the national holiday. But the

Boomerangs, Green Deltas, Bounty Hunters and Wingnuts were at work, business as usual. Unable to overcome the lingering disgust of finding a cockroach in the lime Jello months before, I wouldn't go to the mess hall for turkey and dressing, opting instead for a free dinner offered at the O Club that evening.

By November, more than halfway through my tour of duty, the unconscious transformation from fear and naïveté to a battle-hardened mentality was complete. As the possibility of death had changed to a probability, all emotional connection to what we did from day to day had disappeared. Decisions were made in a cold, deliberate fashion based on what each situation required, and I didn't think about my personal safety anymore (most of the time anyway). The fear that was present early on had largely evaporated, an unnecessary element in what seemed like the inevitability of death. In its place was a cynicism that made the giving of thanks nearly impossible. How could a person be grateful when he had been sentenced to this place and this duty by his own government?

Like many of the parts we needed for our equipment, by that Thursday gratitude had been placed on backorder. It seemed impossible to be grateful when good people were dying in a war effort that the folks back home supported less and less every day. By now, even veterans returning from Vietnam and the old war hawk, Walter Cronkite, had declared our daily service and sacrifice a forlorn cause.

Forced by our government into a war we didn't want to fight, having our hands tied by what seemed like insane rules of engagement, despised by at least a portion of the American public for doing our duty, and hated by many Vietnamese as imperialist killers, it seemed like we were men without a country. In this no-man's-land we worked and fought for each other and for valor, but without gratitude. It seemed that thanksgiving too had become a casualty of the war.

--- Ω ---

The "Vietnamization" of the war was the name for the process designed to allow the U.S. to withdraw from Southeast Asia and turn the total responsibility over to the Vietnamese. The process ultimately failed when North Vietnam invaded and took over South Vietnam in 1975. But in the fall of 1970, Vietnamization was proceeding apace as U.S. units retired to the States and left their equipment and supplies for the Vietnamese. Both the 121st and 336th AHCs in Soc Trang had been decommissioned in October and it was just a matter of time until units at Can Tho, Vinh Long and other airfields in the delta followed suit.

This process was the centerpiece of the end game for U.S. forces that had been advising, training and fighting for and with the South Vietnamese for more than a decade. And while it may have looked hopeful from a bird's eye view, those of us who saw ARVN troops and VNAF pilots up close had serious doubts about their competence and their ability to handle the fighting themselves.

Our reservations seemed to be affirmed on 24 November when a VNAF Huey with 14 people on board was involved in a mid-air collision at Binh Thuy AFB with a fixed wing RU-6 (Beaver) from Can Tho Airfield. The Pilot of the RU-6, CWO Robert Perry*, and another crew member, SP6 Norman Evans* were killed in the accident along with all the other people on board the two aircraft, mostly Vietnamese. We shared our hangar with the company the Beaver belonged to, the 156th Aviation Company, so this tragedy was close to home.

--- Ω ---

In addition to the night crew that worked in the hangar, we had a 3-man crew that worked out on the flight line, in the open air of the revetments, taking care of problems that didn't require the choppers to be in the hangar. These three guys were some of our best people. They were enthusiastic and experienced, and seemed like they could fix just about anything with safety wire and a pair of pliers. I loved working with them.

The night of 27 November, we had a stiff cold wind across the airfield, with rain that lasted from 8 PM until 2 AM. They had their hands full replacing the main rotor head and blades on a Huey out on the line. But in spite of those terrible working conditions, they got it done. And after that little chore was finished I worked with them to fix a high frequency vibration problem in the tail section of another Huey. High frequency vibrations, which were usually felt by the pilots in the seat of their pants, were extremely hard to track down. Most of the time the only way to solve them was by changing out parts until the problem went away. This time the problem was solved with a new tail rotor hydraulic unit, the first part we replaced, based on an educated guess.

That line crew was remarkable. No matter what the problem was, they always got it fixed. And that's one of the reasons the 191st had one of the highest aircraft availability rates in all of South Vietnam!

**OER #2**

When I was reassigned to the Maintenance Platoon, it was Henry Solomon's responsibility to rate my performance for the 38 days I had been under his command. On 28 November he completed my efficiency report and handed it to Major Simon to indorse, which followed the Army's two-step rating process.

Based on my run-in with Simon several months earlier, I expected his evaluation to be negative, but I was surprised to find both Solomon and Simon had given me high marks. An excerpt from Simon's narrative read, "As an aviator 1LT Barth has continuously set a fine example for new aviators assigned to his unit both in professional flying and tactical operations." It was clear from this comment that Simon did not hold a grudge for my outburst, and I was surprised at his willingness to overlook something as personal as that incident had been.

# Chapter 8 - DECEMBER 1970

*My Sweet Lord* - George Harrison

When I was a Boomerang in the 1st Platoon, from mid-May through the end of September 1970, my attention had been focused on learning how to fly a helicopter in combat; adapting to the culture of the Army in a foreign country; and getting to know the personalities in the 191st. By December I had been a Wingnut for two months and my attention had turned from things outside a Huey, to the nuts and bolts of its internal operation.

After a couple of months, I'd come to realize that maintenance was the life blood of the company. We Wingnuts quite literally had our hands all over every aircraft in the company, and we knew that each one had its idiosyncrasies. Our mechanics knew this as well as we pilots, and over time we all grew to expect certain Hueys to behave in certain ways. We knew every aircraft from the inside out, but not as well as some others did.

Every Huey had a technician, known as a crew engineer or crew chief, who was responsible for maintaining the chopper in good working order. To do that required a wide range of knowledge, some of which was acquired through formal training, and a lot that was acquired on-the-job. The enlisted men who held these positions were well motivated to do a good job because they also doubled as a door gunner on every flight their choppers made. If their maintenance was sloppy, they could pay the price for it along with the rest of the four-man crew.

Maintenance work was often required somewhere other than the hangar or flight line at Can Tho, like the morning after the ambush at Rach Soi Airfield in June when our two Hueys had to be lifted back to Can Tho to repair battle damage. At other times, Hueys might go down in a rice paddy, or a lake, or a canal, and someone would have to go in and recover them. Still other times a pilot might shut his bird down because he got a warning light of some kind, and some mechanic would have to fix the problem.

**The Wingnut Flight Crew**

Wingnut 6 was technically Diamond's call sign, but the full Wingnut flight crew included me, the crew chief, SP6 Burt Hornblower, and SP4 Thomas Smiley, the door gunner. Hornblower was a very competent and reliable mechanic who had been handpicked by Don Jefferson to be his crew chief. He was tall, slim, dark haired and mild mannered, and instant to obey an order ... an unusual and highly desirable trait in a GI of that day and place. Having a strong desire to get home, he kept his equipment in superb condition. His M-60 was always inspection-ready, our helicopter was nearly spotless, and it was always stocked with frequently used replacement parts. He was a top notch crew chief, and cared about the rest of the crew as much as he cared about himself.

SP4 Smiley was withdrawn, quiet, and rarely spoke. He was of medium height and build with blond hair and a clear, Scandinavian complexion. He was neither a self-starter nor a confident mechanic, so he followed Hornblower's directions to the letter. They got along well with each other and with me and

Diamond, and collectively we were an efficient and effective crew.

And efficient we certainly needed to be, because we were on standby 24/7 to recover or repair all 13th Battalion choppers in the field, no matter where they were. Diamond approached this responsibility vigorously, setting a ambitious goal for us to be "on station" for a recovery within an hour after we were called. For the first 12 recoveries we were called out for in December, we achieved that goal, which proved it could be done.

**Destiny**

Back in June 1967, a new restaurant opened on Kossuth Street in a neighborhood on the southeast side of downtown Columbus, Ohio called German Village. Schmidt's Sausage Haus was owned by George and Grover Schmidt, and I was one of the original employees when it opened, along with my wife who had pointed out the opportunity for employment to me. I worked at Schmidt's part-time as a bartender, cook and then maitre d' from the time it opened until I graduated in March 1969.

About a year after I helped open Schmidt's Sausage Haus in Columbus, another OSU student named Rip Hardy joined the Schmidt's crew. Rip was a native of central Ohio and a ROTC officer like me. He was smart, reliable, friendly and honest, and because of my experience I was asked to teach him the ropes at the restaurant. We didn't have time to get to know each other very well, but he was an eager learner and we got along well on the job as Rip picked up the details quickly.

When I left for active duty in March of '69, we didn't expect to see each other again. But, incredibly, he showed up in the 191st during the second week in December as a green 1LT straight out of flight school and AMOC. It was amazing that Hardy showed up at Can Tho. With dozens of helicopter units in Vietnam and hundreds of pilots to be assigned to them, what were the chances I'd see him there?

I figured this reunion was not coincidence; it was more like destiny. Just as Rusty Miller and Al Hoppes had been the friendly faces there for me when I arrived, I was now in that role for Hardy. And in the near future, we would be sharing many hours and many experiences as I played the role of more than just a friendly face.

**Hooks, Cobras and Loaches**

CH-47s, known as Chinooks and made by the Boeing Company, were first introduced in 1962 to fill a need for a helicopter that could lift heavy loads in combat conditions. Powered by two jet engines, its twin rotors turned in opposite directions which eliminated the need for a tail rotor and provided better control when lifting unbalanced loads. The Chinook (also referred to by pilots as a "hook", or more condescendingly, a "shit hook") could carry troops or loads internally, as well as by sling load below the helicopter. Because of their high cost, vulnerability to enemy fire, and relative rarity, they almost never ventured into situations that were not secure.

AH-1G gunships, also called Cobras or "Snakes," were powerful aircraft introduced in 1967 for Army use. They were sleek and highly effective in providing intense amounts of concentrated firepower with their 2.75" rockets, multi-barrel minigun (firing 7.62 mm rounds) and 40 mm grenade launcher. The two-man crew of a Cobra was comprised of a pilot and a copilot/gunner, who flew in tandem (i.e. one behind the other) in contrast to Huey pilots who flew sitting beside each other. The tandem seating was necessary because the Cobra's fuselage was no wider than three feet.

Two Cobras working in conjunction with two Loaches (OH-6 Low Observation Helicopters) made very effective hunter-killer teams, good at locating and destroying concealed targets. From the perspective of a slick pilot delivering troops to an LZ, Cobras were less desirable than the UH-1C (Charlie model) gunships that the Bounty Hunters flew because they flew too fast. Huey slicks flew at 80 knots into LZs, as did Charlie model gunships, which could stay with the slick flight all the way to the touchdown point and provide sustained fire cover while troops exited the aircraft. Likewise they stayed with the flight leaving an LZ until they were out of enemy range. Cobras on the other hand, diving toward an LZ at 140 knots or more, could not effectively stay with a slick flight and would make several diving runs as the slicks descended to an LZ. That left long moments when the slicks had no gunship cover at all - a very undesirable situation.

**Whirlwind**

The Wingnut flight crew could pretty well handle minor repairs of helicopters out in the field, thanks to Hornblower's mechanical abilities and his fantastic supply of spare parts. But if we were called out and found the aircraft could not fly and could not be repaired, there were other essential support assets to call on for help. One of them was the 271st Assault Support Helicopter Company at Can Tho (the Innkeepers) that provided heavy lifting capability using twin rotor CH-47s. Without them, we would have been unable to provide the support the battalion needed when aircraft were out in the boondocks and unable to fly.

On many occasions we were called out to retrieve helicopters that could no longer fly. Sometimes they were at safe locations like an airfield. At other times when they had been shot down during a combat mission, they were in much less secure conditions. In either case, our procedure was for Diamond and Hornblower to handle the actual recovery while Smiley and I provided cover for them, me holding our aircraft at a hover nearby while Smiley manned the M-60 to provide covering fire. As a result, I had never worked up close to a Chinook like Diamond and Hornblower had.

In early December, a Cobra gunship pilot shut down his bird at Ca Mau Airfield because of a warning light that would not go out. We were called out to repair it, but when we arrived it proved to be beyond the limits of either field maintenance or safe flight, so the Innkeepers were called to hook the bird back to its airfield.

By the time we arrived, the air traffic at Ca Mau had disappeared. Needing to look at another downed bird someplace else, Diamond left me to handle the Cobra by myself. I tied the rotor blades down and rigged the heavy canvass sling around the rotor head, then sat on top of the Cobra's cowling about 12 feet off the ground waiting for the Innkeeper Chinook to arrive. In the mid-afternoon it was blistering hot sitting there in the full sun, and with very little wind it felt like an oven.

The Cobra pilot who had shut down this bird had parked it on the edge of the runway between the PSP and a water-filled ditch. It sat in a dry space about 10 feet wide that was covered with sand and small stones. Scattered around the area was an empty wooden box and some loose pieces of wood and paper that I didn't pay much attention to, being preoccupied with the heat and the approaching Chinook.

The Innkeepers entered the traffic pattern, landed, and hovered in my direction. I motioned to the pilot indicating I was ready to hook up the Cobra, and he began moving slowly toward me and the Snake. As he got nearer, the downwash from his twin rotors picked up all the loose debris in the area and flung it randomly in all directions. When the outer edge of his downwash started to reach me, the Cobra started to shake and buck from the turbulence, and I could feel it inching across the ground.

The roar of the Hook's engines and downwash was deafening as the Innkeeper came closer and the turbulence increased. I pressed my knees against the cowling to hold on while sand and small stones began pelting me all over. I had no cap or helmet and sand was blowing in my hair and eyes, which I tried shading with my hands. But it was impossible as the whirlwind seemed like it was coming from every direction. Loose pieces of wood and trash were scooped up and swirled in the vortex, and I began to worry that something larger than sand or pebbles would hit me. I felt like I was in over my head!

The Chinook was about 20 feet above the ground, and as it advanced toward me its shade blotted out the sunshine bringing immediate relief from the scalding sun. I looked up to search for the belly hook, squinting to keep sand out of my eyes. The belly of the Chinook was bobbing up and down in the turbulence and the sand in my eyes made it tough to focus, but I could just see the hook through the sand and tears that were now flowing down my face.

I held the sling loops up with my left hand and reached for the belly hook with my right, just as the empty wooden box I had seen a couple of minutes before whirled up into the air and struck me in the side of the head. I automatically ducked as a result of the strike and let go of the belly hook, but I didn't drop the sling loops.

Hovering directly over me now, the Chinook began to sink, closing the gap between me and the belly hook to a foot. Then it quickly reversed, jerking upward a couple of feet, indicating the pilot was having a hard time holding a steady hover. Worried that I could be crushed in this space between the Chinook and the Cobra, I lurched upward and grabbed the belly hook, slipping the sling loops over it at the same time. Then half blind from the sand in my eyes, I slid down off the cowling. My feet landed on the rocket pod mount, and I leaped down to the ground moving away from the Snake until the Chinook pilot could see me. Then I gave him a thumbs-up sign to tell him the Cobra was ready to go. He returned the sign and I moved clear of the tornado of sand, pebbles, boards, paper and other debris that continued to swirl about the Cobra.

The Chinook rose gently, carefully removing the slack from the sling, then with a burst of power it quickly pulled the Cobra straight up into the air. The pilot immediately took a heading north toward Can Tho and the Cobra, momentarily perpendicular to the direction of the Chinook's flight, oscillated awkwardly in the wind seeking to align with the Chinook's flight path. As the pilot gently gained forward movement the Cobra quickly turned into the direction of flight as its fuselage streamlined in the wind.

I wiped my eyes as I watched the Hook and Snake gradually disappear to the north. My thighs and backside hurt from sitting on the Cobra's cowling. My head was sore where the wooden box had hit it, reminding me to wear a helmet the next time. But no discomfort could overshadow the excitement of this experience. It ranked right up there with the adrenaline rush from combat assaults, rescues of downed comrades, being ambushed at Rach Soi, and test flights alone at night. I wanted to carve a notch in the handle of my .38!

**It Pays to Know the Right People**

Robert McCoy transferred out of the Wingnuts, and although I felt unprepared for the task, I was promoted to Assistant Platoon Leader on 5 December. I envied the formal maintenance training both Diamond and his predecessor had. What they learned in the classroom I had to learn on-the-job, and I knew I couldn't possibly learn it all. That meant I had to rely on Diamond Jim for the knowledge I was missing.

Now Jim, being a Texan, had a tendency to exaggerate a little from time to time. So when he told me his inner feelings and plans while we were flying one day, I didn't believe him. He told me that he hated the war and was making plans to get out as quickly as possible. So I had better be prepared to take over his position when that happened.

This news seemed like wishful thinking that lacked any serious possibility. Playing along with him I asked just exactly how he planned to pull off an early return to the States. With his usual broad grin, he said his wife's family had influence with a congressman who was making arrangements for him to be transferred to stateside duty. I considered this story to be purely hypothetical and told him so. He just grinned and said, "Barth, I'm warning you, this is going to happen and you'd better be ready."

I didn't know what to think about this story. No one I knew, or had known, had any clout with politicians, and it was inconceivable that this swaggering southwesterner had it either. But he had warned me about what might happen and wanted me to know that I had to come up the learning curve as quickly as possible. In spite of his story, I was doubtful of his ability to pull it off. It wouldn't be long before the truth was known.

**The Southern Assault**

The least secure part of the Mekong Delta was the swampy forest area along the west coast running south from Rach Gia to the southernmost tip of South Vietnam. An Xuyen (*on zwee-yen*) Province was a wild place where dry land blended into swampland, and then gradually into the open ocean. Contained within its perimeter was the U Minh Forest that provided cover for VC who used the area to infiltrate war supplies and troops. To secure the Delta, the VC had to be cleared from this area.

Ca Mau (*kah mah-oo*) city, also known as Quan Long (*kwahn long*), was about 80 miles southwest of Can Tho Airfield, centrally located in the upper An Xuyen Province and the Ca Mau peninsula. Its population now is about 200,000, so the population in 1970 was, no doubt, significantly less.

In late November, and continuing on into December and subsequent months, the 164th Aviation Group and the 13th Combat Aviation Battalion provided support operations for the ARVN ground forces attempting to clear the province. Both the 191st and 162nd were involved in the operations and Boomerangs were in the air every day over Ca Mau, including the Wingnut 6 crew.

As the operation ramped up, the demands on our time increased substantially. Ordinarily we handled one or two callouts for maintenance a day, but by then we were averaging more than four a day, along with the work needed to keep the 191st choppers in the air. Most days we were off the ground and en route to the first mission by 6 or 7 AM and finishing up the last maintenance callout by 7 or 8 PM. Then it was back to Can Tho to do a test flight or two before finding something to eat, along with the necessary daily ration of booze.

Feeling alone and deserted because Diamond and I were so often away on maintenance callouts, Dex Woodley was frustrated that most of the humdrum maintenance tasks had fallen on his and Rip Hardy's shoulders. While he would complain vociferously to me, I gathered that he never said anything to Diamond.

Day after day the area south of Ca Mau was assaulted by U.S. and ARVN forces, and our maintenance callouts kept pace. On the 17th we were called to recover a Huey that had been shot down in a swamp south of Ca Mau that was not secure. This was a new wrinkle for us because we usually recovered complete Hueys after the area had been secured. But for some reason unknown to us, and to our dismay, Guardian 6 called for an immediate recovery before the area had been fully secured.

We took off from CTA and got map coordinates of the location. En route, Diamond and I discussed how to handle the mission. He wanted me and Hornblower to rig it for hooking out, but I felt uncomfortable with the situation and told him so. He then insisted that I handle this on the basis of needing the experience in preparation for his eventual absence, but I argued that the circumstances dictated that the two most experienced should rig the ship for pickup, and that was him and Hornblower.

Our debate paused when we reached the area of the downed Huey and saw it was sitting in water 4-5 feet deep. Just getting to it would be difficult because we couldn't approach closer than 25-30 yards from the Huey. Whichever one of us went, he would have a tough slog through a lot of deep water just to reach it.

Jim was flying as we circled the downed Huey 1,500 feet below us, figuring out the wind's direction and selecting a place where we could hover close to the downed bird. Landing was out of the question because of the water's depth. A long hover above the water would be necessary while the Huey was rigged for its lift out. I didn't feel confident about handling a sling rig job like this one, and became more insistent that Diamond and Hornblower handle it while Smiley and I waited. Looking me directly in the eye Diamond said, "OK, Barth, but you owe me one!"

The landing approach was set up and I took control of our Huey while Diamond and Hornblower tore off their flight helmets and gloves and prepared to jump into water of unknown depth. I brought us to a hover with the skids dipping into the water as Diamond opened his cockpit door and jumped out into chest-deep, muddy water. Hornblower threw the rigging gear to him and then jumped out on the left side as well, and I watched as they struggled toward the downed Huey, half swimming and half walking through the water.

The water-covered field where we were was 60 or 70 yards across, with a tree line about 30 yards to our right front. The downed Huey was about 30 yards to our left. Over the intercom I told Smiley to keep an eye on the tree line and watch for enemy fire that might come from it, and return the fire if he saw any. Then I turned my attention back to Diamond and Hornblower.

Their progress through the water was slow and I watched in frustration trying to think of something I could do to speed up their progress. Several gunships had materialized out of thin air and started shooting up the tree line to our right front, a most welcome sight. In the process, a Chinook appeared overhead and started to circle, waiting to make a hook-up and drag the Huey home where it could be repaired.

They reached the crippled Huey and climbed onto the roof to rig the sling around the rotor head. I continued to hover in place hoping to get out of there in one piece, my attention split between the gunships working the tree lines and our two Wingnuts on the roof of the Huey, working the rigging. The whine of our engine was loud in the background, but above it I heard two muffled "whumps" that registered on me as mortar rounds hitting nearby. I turned the Huey to the right, toward where I thought the sounds had come from, and asked Smiley if he had heard anything like mortars exploding, but he said no. So I concluded the sounds may have been rocket rounds from the gunships circling overhead, but I was not entirely sure.

A couple more minutes went by before Diamond waved the Chinook in for the hook-up. The big twin-rotor's downwash was whipping a mist of water up over, and then down through its rotor blades in a continual double cascade that caught the sunlight and produced beautiful, multiple rainbows that stood out against the dark green foliage and blue sky in the background. When the hook-up was complete, Hornblower was already in the water and struggling to get back. Diamond slid off the Huey's roof and began to follow him, his hands and elbows flailing as he tried to speed his progress to get away from the Huey and the cascading water.

In one slow, powerful move the Innkeeper pilot pulled the sling taut and hoisted the Huey. Water rushed out the open cargo doors like rivers. As it rose, the cascade of mist diminished and the rainbows quickly disappeared, and turning slowly the Hook gained altitude and speed and was off toward Ca Mau Airfield where the sling rigging could be re-secured before the longer flight to the Huey's home base.

In a couple of minutes Hornblower and Diamond reached us, and I dipped the skids deep into the water so they could climb aboard. Water poured out of their flight suits as they climbed in, like it had out of the Huey's cargo compartment. They were out of breath and exhausted from their effort as I turned our chopper into the wind and took off for Can Tho. Diamond Jim had that irrepressible Texas grin on his face, and I knew all was well from his point of view.

A month later, Diamond wrote a recommendation for awards for all the members of the Wingnut 6 crew for that 17 December mission. I received an Air Medal for it, which I didn't think was deserved because this was my job. We all had done what was required, as a team.

--- Ω ---

Ignatius Coleman was the 5'10" son of a coal miner from western Pennsylvania. He was good looking, almost to the point of handsome, in spite of his portly frame and round face. He had been the executive officer of the 13th Battalion and was reported to be mean and tough, and on 21 December took the command of the 191st.

It was a welcome relief from Simple Simon, a man almost no one respected. But we didn't know what to expect from this new CO. Over the next five months I would learn to appreciate his straightforward, no-nonsense approach, and the way he cared about the pilots in his command.

--- Ω ---

In seven months, Al and I had transformed our 7'x10' hootch into something that resembled a home. Along the north wall, which was on the left as the hootch was entered, was a small refrigerator, a small desk that we cut a hole in the top of to house a small record turntable, a TV set that received a couple of Vietnamese stations and an AFVN station, stereo speakers, and a set of shelves with plywood doors.

Along the east wall (the wall opposite the door) there was room for one folding chair and two built-in closets where we kept our meager clothing and other supplies. The metal bunk beds were opposite the closets along the west wall. Between the bunks and the door on the west wall there was room for a shelf which was generally a catchall, but it became the focal point for Christmas. On this shelf we placed a small Christmas tree that my wife had sent, along with some holiday candies, candles and cookies. It also held holiday treats and gifts from family members and various organizations like the USO and the Perry Packin' Gals.

It was nice to be reminded again that people back home had not forgotten us.

**Holly Holy**

I didn't fly on the $24^{th}$, but I did make a telephone call to my wife to wish her a Merry Christmas. When she came on the phone and I heard her voice, I choked and could only say, "I love you," words so special at that time of year. They were the only words I knew of that could drown out the incessant song of war in the background, that could heal a heart, that could restore the world to a higher humanity. They weren't said enough.

An operational stand down was promised to us on Christmas Day, so early on Christmas Eve several of us got together to celebrate the Advent, our holiest night. My bunk mate went off to the officers club for a party, so Tom Yost, Mike Tschudi and I got together in my tiny hootch to talk, smoke Marlboros, drink wine and listen to music. It was always special when we three got together, because our extended conversations were rich, often deeply reflective. This night would be no different.

Tom Yost was a special friend, but Mike Tschudi was special and unusual. He had been a seminarian, and he loved the Catholic Church. But his doubts about his ability to make the commitments required of the priesthood caused him to drop out of school, and he was drafted into the infantry and shipped to a rifle platoon in Vietnam. In the space of only a couple of months he'd gone from a saintly environment to its extreme opposite, and I couldn't imagine how shocking that transition must have been.

But Mike was not just priest material, he was a gentle, caring soul who wrestled with angels, both good and bad. At times his internal struggles were torturous, emotional conflicts that left him exhausted from the metaphysical battle. But they also strengthened his character and opened his heart to others, as shown in the words he wrote to me that Christmas:

> *I couldn't put my finger on that certain something which lets us talk till 2 A.M. or sit for hours in the silence of shared unspoken thoughts*
> *I know only that far too few can do it and have it mean something*
> *And I say Thanks for being one of those few*
> *Best always - Mike*

When we got together that night the mood was light and festive at the outset, but as the evening lengthened the atmosphere turned somber. Eventually we started listening to the songs on a Neil Diamond record that someone had dropped off at the hootch. Many of the tunes were familiar and we could sing along with them, and by the time the record got to *Holly Holy*, the wine and the talk had done its work; we were entranced.

As the song began we irresistibly rose, put our arms about each other's shoulders, swayed to the music, and sang along.

| *Holly Holy* - Neil Diamond |
| --- |

Later it was agreed that that moment was indeed Christmas for all three of us.

Even now, more than forty years later, when I hear *Holly Holy* it immediately takes me back to that moment in 1970 when the unending thoughts of war were interrupted by a song of promise and healing. It was a song that touched the war weary, the man who could not walk upright; and that lame man flew. In no uncertain terms the better angel of our existence had said, "I love you." And we all flew.

--- Ω ---

Christmas morning I was disappointed to awake and find the war was still on! I knew the workload would be light that day because everyone else was off duty. I arrived at the hangar at noon, did a test flight on one of our M Model (Mike model) gunships and then called it quits too.

**Mike Models**

Back in August, the 191st had received several M (or Mike) model Hueys, which were C (Charlie) model gunships that had been modified with more powerful engines (1,400 shaft horsepower instead of 1,100). There were several versions of this essentially experimental aircraft, and they were assigned to only two companies in Vietnam, the 191st and a unit in the 101st Airmobile Division in I Corps. Our version, designed to operate at night, was equipped with miniguns on each side of the fuselage, standard 2.75" rockets, and an infrared camera system (known as an ENFANT system) mounted on the nose. Each pilot had a small TV monitor in front of him in the cockpit to see the images produced by the infrared cameras.

The tactical concept was for the Mike models to fly together at night in blackout conditions, i.e. with no external lights on at all, and only red cockpit lights, so that VC could not see a target to shoot at. By operating in free-fire areas that were provided by the province operations offices, ground targets identified as infrared heat sources were assumed to be enemy soldiers that could be taken out. The concept was logical, but the successful execution of such operations depended on perfect communication between friendly ground forces, civilians, operations centers and the 191st pilots.

The combination of at least one bad assumption, the inability of pilots to confirm targets visually, and communication SNAFUs produced conditions that resulted in several atrocious errors involving the Mike models while I was there. (SNAFU is an acronym for Situation Normal, All Fouled Up, a common Army term.) One error was the attacking and destroying of a friendly ARVN patrol that the operations center did not know was in the area. Another was the destruction of civilians in sampans just off the coast that had wandered into the free-fire zone. In both instances the pilots were absolved of responsibility.

These incidents furthered a growing belief of the insanity of daily operations in Vietnam. It was not logical to me that the threat from VC attacks was significant enough to justify the extreme assumptions that Mike model operations required. Later, back in the States at Ft. Riley, Kansas, my 1st sergeant would often say, "Assumptions are the mother of all f---ups." The Mike model failures I knew about were deadly confirmations of that observation.

--- Ω ---

VNAF participation in the war was growing daily, and so were their losses. For the last couple of months we had been turning over more and more of our Hueys to VNAF units to meet their needs for additional Hueys, but our stock of aircraft that met their standards was growing thin. On the 28th we got orders to turn over the last two Hueys in the company that met their requirements, and one of those was the Wingnut 6 Huey, the aircraft that Hornblower and Smiley had kept in immaculate condition. It was hard to let it go, but let go we did, taking an older Huey from another platoon in order to continue our recovery missions.

--- Ω ---

On 27 December something happened that earned me an Air Medal for heroism. The citation for the medal indicated this, but in complete honesty I cannot remember this incident.

**Magnum**

Charles (don't call me Charlie) was a relatively new pilot in the company. At about 5'9" tall with a slim build, a slow southern drawl and an equally slow manner, he seemed completely out of place in both the Army and Vietnam. It seemed as though he should be hobnobbing with southern aristocracy and drinking Southern Comfort, or Mint Juleps. But as a Green Delta pilot his smooth southern style was tailor-made for hauling the brass around the delta.

Between Christmas and New Year's there was a floor show one night at the O Club, so Charles and I went together to see it. In preparation for the show, and in keeping with the general festivity of the season, Charles had bought a magnum of champagne and was determined to consume it that evening, all by himself.

When we reached the O Club we grabbed a good table out on the patio, and when the show started Charles popped the cork and guzzled champagne until the bottle was dry. Even stretched over several hours, this was an astonishing feat, and by the end of the show he was three sheets to the wind. As the crowd dispersed, he looked at me across the table, his eyes bloodshot and crossed, and garbled he was going to smash the empty bottle on the concrete patio floor. I thought he had turned into the Mad Hatter from *Alice in Wonderland*, and hoped to leave before something stupid happened.

A magnum bottle was made of very thick glass, and I knew if Charles was successful there would be a tremendous explosion of glass in all directions. So as he stood and raised the bottle above his head, I pushed my chair away from the table to get out of the way. Letting the bottle go from head height it fell to the floor in front of him with a dull thud. It didn't break; it didn't even throw a chip.

Looking down at the still-intact bottle, he realized he had to throw it harder if it was going to break. So he reached down and grasped the bottle by the neck, and threw it hard onto the concrete. I expected the bottle to break, but this time it hit the floor and bounced nearly straight up, striking Charles squarely in the forehead and momentarily knocking him cold. Both the bottle and Charles hit the floor about the same time, but the bottle was unharmed.

A few seconds later he came to, rubbing his forehead and complaining about the sore welt on it. I helped him get to his feet, and got him out of the club while he complained that we were leaving his bottle behind. As we left, I imagined Charles at a party of proper southern socialites, politely sipping champagne cocktails-- and then smashing the bottle on the floor! It might be OK at the CTA Officers Club, but not back in Georgia.

The next day Charles had a huge black and blue goose egg in the middle of his forehead that his cap was too small to hide. And when asked how he got it, he replied in his slow drawl that he was pleading the 5th!

**Setting a Record**

As December's end came near, we stayed busy with recovery missions in support of the southern assault. By the end of the month we had tallied 37 aircraft recoveries in a total of 26 days, exceeding the previous monthly company record by 15. We were proud to have achieved it, and knew we owed it to Diamond Jim and his aggressiveness.

--- Ω ---

Because I had worked on Christmas Day I had the night off on New Year's Eve. So late that afternoon Jack Wyatt picked up Tom and me and flew us to Vinh Long Airfield to welcome in the New Year with him and his buddies. We spent much of the evening in the O Club watching a floor show. Afterward we watched another show as a drunken pilot demolished a group of banana trees near the pilots' billets, using the tender plants as tackling dummies. When he finally collapsed from exhaustion the place looked like a tornado had gone through leaving banana tree leaves all over the place. It was a mess.

Near midnight the guards on the perimeter of the airfield began firing their rifles and machine guns, and launched flares into the sky to celebrate the turn of the calendar. I wondered what they would have done after wasting their ammunition that way had the VC attacked the airfield just a few minutes after midnight. Fortunately that didn't happen.

# Chapter 9 - JANUARY 1971

| *New World in the Mornin'* - Roger Whittaker |
| --- |

Wyatt dropped us off at CTA late in the morning on New Year's Day. It was not the new world in the morning I hoped for; it was back to work as usual.

An order had recently been issued directing that the tops of the horizontal stabilizers on all Hueys were to be painted orange, and this was the day to get it done. Just about everyone was busy spreading ugly orange paint over the beautiful olive drab of their Hueys, transforming the uniformity I liked into an ugly mess.

Later that day the Rose Bowl game between Ohio State and Stanford was scheduled to be on AFVN radio. I thought there was no way a group of Stanford elites with a record of 8 wins and 3 losses could beat a team of tough Midwesterners like the Buckeyes, who were undefeated.

But I was wrong. Jim Plunkett, the Stanford quarterback and Heisman Trophy winner, diced the Ohio State defense with his passing and drove the Indians to a convincing whipping of the Buckeyes, 27-17. This was essentially the same Ohio State team that had been national champions just two years before that. It was a disappointing end to an otherwise fine season of college football, followed by this fan from long distance.

**The Ditch Jumper**

Maintenance recovery missions took the Wingnut 6 crew all over the delta, and in the process we occasionally dropped things off at outposts. One day we visited an artillery unit that was located next to a banana plantation to deliver papers in a large sealed envelope. We found the plantation easily enough and saw that it was laid out with trees planted in 12 foot wide rows, separated by eight foot wide ditches of water that resulted when dirt was excavated to create the tree rows. One of the rows was nearly empty and that provided a good landing spot, even though it was a fair hike from the unit's headquarters.

Diamond put the chopper down there and I volunteered to deliver the envelope while the rest of the crew relaxed. When I stepped out of the Huey it looked like a short cut could be made by jumping across the ditches. However, I couldn't leap the eight foot width from a standing position. Looking down into the water, I could see leaves and grass lining the bottom of the ditch, which was about a foot deep and much shallower than I had anticipated. It seemed clear I could cross it by planting one foot in the shallow water in the middle of the ditch, and then the other foot on the other bank.

I closed my eyes and envisioned what I had to do, imagining I would get a wet foot when my left boot hit the water. I took a step back anticipating what the bottom of the ditch would feel like under my foot, got a short running start, and leaped! My foot hit the water, and searching for solid footing it continued below the anticipated bottom of the ditch, followed by my entire body. I grabbed a breath of air before my head went under, leaving my baseball cap floating on the surface as the only evidence that anything had happened.

It was a huge shock, of course, and a moment later I spluttered to the surface, blowing away the water that was running off a banana leaf draped over my head. Diamond, Hornblower and Smiley were still in the chopper and convulsed in laughter as I clawed my way out of the ditch and retrieved my cap. Water was dripping from my flight suit and the envelope that was still in my hand, and every time one of them glanced at my bedraggled condition his guffawing renewed. It must have been the funniest thing they had ever seen. I couldn't help but laugh along with them.

When I finally delivered the envelope to the artillery commander a few minutes later, he was not happy it had been soaked. I never found out if the contents were still readable.

**Fast Eddie**

1LT Edward Johnson was about 5'10" tall with sandy hair and a Mona Lisa smile that made others wonder what he knew that they didn't. He had joined the company a few months earlier and flew with the Green Deltas, but when he wasn't flying he was the technical supply officer (TSO) for the company. Johnson was the guy who made sure we had the parts we needed to keep the Hueys in the air, and everyone knew it was better not to ask any questions about how he did what he did.

Back on 12 November when the excess parts were discovered, squirreled away in the 162nd's hangar, the feces that flowed downhill to save Simon's behind were caught by Eddie. There was an investigation, of course. (There were many investigations in Vietnam because the military service was loaded with graft and corruption from top to bottom). But it was so complicated that a resolution was still pending by January (and the investigation was still active when I left the 191st in May, four months later).

Johnson was quiet by nature. In the hours I spent with him at the O Club and the Boomer Lounge, he almost never talked openly about anything. And about the excess parts affair he was very tight-lipped. Maybe it was because he was told not to talk about the incident while the investigation was going on. But whatever the reason, his reticence to talk and the mystery that behavior spawned prompted someone to start calling him Fast Eddie, and the nickname stuck.

Anyone who was around for any length of time knew the Army's supply system could be challenging. On one hand, regulations allowed only so many of each kind of part to be kept in inventory. On the other hand, the parts that were actually needed were the result of what was going on in the war, down at the combat unit level. Often the demand for certain parts exceeded the supply and no one wanted to take the heat that came from not having flyable choppers because of parts shortages. So any supply officer worth his salt would hoard or trade parts. It just came with the territory.

I thought Johnson was simply a victim of circumstances and was in the wrong place at the wrong time back in November. It was obvious on the surface that he had not been in the company long enough to have created the excess inventory, and was getting a bad deal by being held accountable for them. Fast Eddie was not really "fast." He was just paying the price for what someone else had done trying to keep our choppers in the air, and was probably another casualty of the times and circumstances.

**Hawaii**

On Monday, 18 January, I left Can Tho for rest and recuperation (R&R). I knew the Maintenance Platoon would be short handed with only three pilots, but was certain they would be OK without me for a week. And by then I was eager to get out of Vietnam.

I caught an Air Force flight from Binh Thuy to Tan Son Nhut, and another flight to Guam, and finally one to Honolulu. I was looking forward to seeing my wife after being apart for more than eight months, but when we finally saw each other it was awkward. It seemed like I didn't know how to act, or what to say anymore to an American civilian. Having grown used to thinking in the language of war, I felt out of place in a civilian society that didn't understand what I and my comrades had seen and done.

After our rendezvous, we spent a couple of nights in Honolulu at the Outrigger Hotel and visited the USS Arizona Memorial at Pearl Harbor, but the experience didn't register on me emotionally. We took a chartered flight around the islands taking note of the nearly endless pineapple plantations, and all the while I was thinking how boring it was to fly when no one was shooting at us. I told the pilot I was a Huey pilot with fixed wing flight time, and asked if he would let me handle the controls of the twin engine Cessna 411, but he said no. (In retaliation I secretly vowed not to let him have the controls of my Huey in the slim likelihood that he would visit Can Tho!)

We landed at the airport on Maui and took a cab to the Sheraton where we had reservations. The hotel was beautiful, perched on a huge black outcropping of lava rock next to the Pacific, with the beach spread out below. It was picture perfect.

The next several days on Maui were spent sightseeing, but the sights were of little interest to me. I enjoyed my wife's company and tried to be hospitable, but I couldn't relax knowing I still had four months of duty left in Vietnam. To some degree we were victims of too much time away from each other, and my survival mindset cut me off from any sort of real vulnerability.

One sunny morning we went down to the beach to go snorkeling. As I stood in the surf flushing the face mask with water, an older man came up to me and initiated a conversation. He asked me what I was doing in Maui. I told him I was on R&R from Vietnam, whereupon his demeanor changed completely. With a furrowed brow and a snarl on his lips he tore into me verbally like a vicious dog. The Vietnam War, so he said, was nothing in comparison to the difficulties of battle in World War II, and troops in Vietnam couldn't hold a candle to those veterans. He went on, saying the U.S. was losing the war and it was the fault of people like me who didn't know how to fight, and were essentially cowards.

His attack amounted to an ambush, and being totally unprepared for it, I was dumbstruck. Not knowing how to respond to his tirade, I turned away and went back to the beach, disturbed by the short encounter.

Over the years I have replayed this incident in my mind trying to figure out how I could have held my own, and stayed with that conversation. In writing this, I know how I should have responded to his attack, but its 40 years too late and I am no longer concerned about what he thought.

But this encounter serves as a good example of the type of abuse Vietnam vets received for doing the duty their own government compelled them to do.

Diamond and Woodley picked me up in Saigon after leaving Hawaii, and when I got back to CTA there was as warm a welcome as anyone ever had. It was most gratifying and I was satisfied to be "home" with people I knew I could count on. The men of the 191st had my back, and in eight months they had become as much my family as anyone else was.

R&R had not been what I expected it to be. Instead of a restful reconnection with my wife and civilian society, it had been a week that provoked mixed feelings, punctuated by a difficult encounter that foreshadowed what my return home might be like. I was now aware that a definite shift in my thinking had occurred and I no longer wanted to deal with the menial and meaningless drama of daily civilian life. From my war-hardened perspective, anything that had less than immediate life or death consequences didn't seem worthy of attention or effort.

In retrospect, I realize that the whole of my R&R experience was a culmination of the previous eight months that represented a major turning point. I no longer saw the Vietnamese people as inferior, but as equal people worthy of deeper respect, deeper consideration, and deeper association. At that moment, however, I didn't realize how much this change would affect the rest of my tour, or the rest of my life.

--- Ω ---

Twenty-six January was the first day of Tet, the Vietnamese Lunar New Year celebration, and the airfield was closed to all Vietnamese. It was a precaution against the possibility of an attack similar to the Tet offensive in 1968 that had taken both the U.S. and South Vietnam governments by surprise. This year, however, it was all quiet.

--- Ω ---

Late the afternoon of the $26^{th}$ we were called out to recover a Cobra. I rigged the head and hooked it up to the Chinook in another thrilling ride atop the cowling. When the Hook picked it up, the pilot accidentally jammed the Cobra's tail stinger into the ground and ripped it out of the tail boom. (The stinger is a 2 foot rod attached to the frame of the tail boom in Hueys and Cobras, and sticks out the back of the tail boom like the stinger on a bee. It's designed to protect the tail boom from damage.)

I knew this would make some CO very angry because they had no tolerance for aircraft being damaged during retrieval. The Hook pilot probably got reamed for this, because every accident had to have a scapegoat in the Army in Vietnam.

# Chapter 10 - FEBRUARY 1971

Nine months down, three to go.

R&R was history and it was back into the fray. I was back with my 191st "family," but the days were now just as lonely as the nights.

| *Lonely Days* - The Bee Gees |
|---|

The assault on the southern delta was going strong and maintenance recoveries were heavy. Almost every day Diamond was clucking at me about him leaving for the States any day, and I had better be prepared to handle the platoon on my own when he was gone, blah, blah, blah. At this stage of the game I really didn't care whether he left or stayed and was tired of hearing about it.

**Tired of Waiting...**

Tuesday, 2 February, Al and I teamed up to fly to Tan Son Nhut AFB to get a tail boom for the IV Corps commander's Huey. At that time Tan Son Nhut was reported to be the second busiest airport in the world, just behind O'Hare Airport in Chicago. But that fact bothered neither of us; we were looking forward to seeing what it was like to fly into an airport busier than Can Tho.

The morning was clear as we took off, but as we got close to Tan Son Nhut there was a lot of ground fog and a thick haze that extended several hundred feet in the air made it impossible to navigate visually. We called the Tan Son Nhut tower and reported we were inbound for landing, but could not navigate due to the foggy conditions. The tower operator offered us a ground controlled approach (GCA), which was a real luxury to everyone in the pilots' community, and we gladly accepted.

A GCA was as close to personal service for pilots as one could get. It involved a radar operator giving specific directions over the radio, such as "turn left 20 degrees," or "begin your descent now." It was almost impossible to screw up if you followed the directions explicitly. The only problem with it was the time it took to give directions to a slow Huey, during which all other air traffic had to be stopped.

We lined up on final approach a good way out from the end of the runway and the GCA began. About five minutes later we broke out of the clouds at about 200 feet above the ground, nicely lined up for a landing on the main runway. Glancing left, I could see some big airplanes on the taxiway waiting for us to land: a C-141, a Northwest Orient 707 and several other large aircraft, all generally fitting the category of "super zoomies" as we helicopter pilots called them, planes that traveled much faster than our 80-knot Huey. We were holding up a lot of traffic with this GCA for our little chopper.

The pilots of those aircraft might have been thinking about the song *Tired of Waiting for You* by the Kinks that was popular in the U.S in 1965.

**Let's Make A Deal**

The missions from battalion were still being piled on the company every day, keeping our hangar full of work. With the shortage of mechanics that had developed in the last couple of months it was impossible to keep up with it all, and we were always hoping for new Hueys to arrive that didn't need much maintenance. So when the 164th group maintenance officer called us the first week in February and said there was a brand new H model waiting to be picked up in Saigon, Diamond and I leaped to the task. We went to Long Binh and picked up a sleek new beauty with not a scratch on it anywhere, including the floor, and it flew like a dream.

The IV Corps commander had a Huey of his own which the 191st maintained for him. It had been in the hangar for almost a week waiting for the tail boom that Al and I picked up on 2 Feb. But that tail boom had to be installed and painted. So, when Diamond Jim and I got back to CTA with the new H model, he proposed to give it to the corps commander if he would agree to let us keep his old Huey once the tail boom work was finished. (His "old" Huey only had 274 hours on it compared to the thousands of hours most of our Hueys had). This offer was too sweet to resist and the corps commander agreed, and with that his not-so-old Huey became the new Wingnut 6 bird. Our new used Huey had less than 300 flight hours on it, and it seemed like a dream to us.

--- Ω ---

By that time the days could not go fast enough, as each passing day meant we were that much closer to going home. But I was also aware that not everyone felt the same, especially the Vietnamese who were already home, and whose tours of military duty would not automatically end with a DEROS. It was impossible for me to know how they viewed this war, but I could imagine that they might be in no hurry for the clock to move as each tick could bring them closer to some undesirable end.

From a letter to my wife dated 6 February 1970:

*Not by the Sun's arithmetic nor my own can I make the days go fast enough. Yet there are those who beg God daily for an extra hour. I wish for them no solitude, no time apart from what they love, and let them have their extra hour.*

*Caught in the Quiet*
*by Rod McKuen*

**China Beach**

One month earlier, back in early January, the company was alerted that Al's Huey, which was full of important VIPs, had had an engine failure during flight and had gone down somewhere in a remote area of the delta. Later that morning we found out that no one had been hurt, nor had the aircraft been damaged.

As Al told the story after his return, he was flying at 1,500 feet when the engine quit with no warning (a possibility that all helicopter pilots dreaded, but one for which we were well trained.) When the engine quit Al's training took over and he autorotated to a safe landing in textbook fashion: he had quickly recognized the engine failure, decreased the power settings immediately, circled to find a clear landing spot, and put the Huey down in a banana plantation without sustaining any damage. It was a picture perfect autorotation and landing that he described in a matter of fact way with no pretense.

Al's handling of the situation was impressive, so the CO rewarded him with a three-day pass to Bangkok, Thailand. It was nice to be recognized and rewarded, but Al was frustrated because he didn't have the money to pay for a trip out of the country. A few days later I went on R&R and the subject was dropped until I got back. By then he had decided that he could afford to go to the China Beach R&R Center near Danang, but he really didn't want to go anywhere by himself.

We talked about the situation for a few minutes, and then he asked me if I would go with him if he could get a pass for me too. I wasn't very enthusiastic about the idea, partly because Danang was 500 miles north of us and traveling there and back would take a big part of the three days, plus I didn't think Diamond would approve of me being gone again. On the other hand, I knew Al wouldn't go alone, so I agreed to approach Diamond about it if Al could somehow arrange to get a pass for me.

China Beach was the nickname for a place known to the Vietnamese as My Khe (*mee kay*). It was close to the DMZ (demilitarized zone) that separated North and South Vietnam. The only thing we knew about it was its reputation as an R&R center. But the name China Beach had a mysterious quality to it that reminded me of Bali Hai in *South Pacific*.

Monday night, 8 February, there was a hail and farewell party for some of the officers who were going home, and as usual there was a lot of drinking late into the night. I talked to Diamond about going along with Al to China Beach, and he said if I could get a pass from the CO it was alright with him. With that piece of the puzzle in place, Al approached MAJ Coleman with the idea, but his response was an emphatic "No!" Al pressed on, however, explaining that Diamond had already agreed to my leaving for three days, but that didn't sway the CO who had his heels dug in on the issue. In a last ditch effort Al asked the XO, CPT Handy, if he would sign travel orders for me and he agreed. It looked like we were on our way to Bali Hai!

Early the next morning Handy signed my orders. We quickly packed a bag with civilian clothes and headed for the flight line to catch whatever flights we could find going north. By early afternoon we were on a C-123 from Binh Thuy AFB to Tan Son Nhut AFB, arriving about 5:30. We had dinner at the O Club and stayed overnight at the transient officers' quarters, then caught a flight on a C-130 to Cam Ranh Bay at 6:00 AM the next morning, followed by a flight to Danang.

Walking down the exit ramp of the C-130 we took to Cam Ranh, I saw a gigantic plane across the airfield, double the size of the B-52 parked next to it. I didn't know what it was, so I asked the Air Force crew chief on our flight. He said, "Oh, that? Why, that's Fat Albert, the aluminum cloud. That's a C-5A." It was the first time I'd seen that illustrious plane, the second largest bird in the world. It was shockingly immense!

From Danang we took an Air Force bus to the R&R center where we got a room that looked out on the beach. Even though we were dog tired from the trip, we hustled down to the water's edge only to find there was no swimming, or surfing, or water sports of any kind because of a strong undertow. All the way from Can Tho we'd been thinking about swimming in the South China Sea, and now we were stopped at the water's edge! It was disappointing, but we were so tired from the trip that going back to the room and sleeping was more attractive than anything else. Instead of relaxing in the sun, we slept at the China Beach hotel. We'd traveled 500 miles just to take a nap!

Later that day we went to the gym where we played handball for a couple of hours, then had dinner at the O Club and watched a couple of movies. The center was well-equipped with an EM Club that had live shows every night, an open air theater, gym, library, PX, a beach house where surfboards could be rented, and an O Club with a stone patio that surrounded the entire building. There on the patio, one could order drinks and dinner, and gaze upon the dark waters of the South China Sea fifty yards away. This was not Bali Hai, but the light of the full moon rising slowly above the distant horizon danced on the peaks of the waves and found their way up a wooden walkway to where we sat. It was another display of the amazing beauty this land had to offer.

The beach itself was white sand up and down the coast as far as the eye could see. The area next to it was equally pleasant with tall palms scattered throughout the R&R Center. The mountains surrounding the area had several large fire-support bases at their summits that were brightly lit at night, giving the whole area a carnival look.

On Friday morning the 12th, we got up at 8:00, ate breakfast and went swimming in the South China Sea. The lifeguard would not let us surf because of the undertow, but the water was cold and as beginners we probably wouldn't have enjoyed it anyway.

Then, all at once it was time to go. We packed our duds and hitched a ride in a military taxi to Danang AFB where we waited two hours for a flight back to Saigon. As we waited, a large group of GIs gathered and boarded an American Airlines 707 for their return trip to the States. They were done with their tours; we envied them and wished we were on the flight too.

After arriving at Tan Son Nhut about 5:30, we ate dinner at the O Club, saw a lousy movie at the theater and went back to the O Club to see a floor show that was standing room only. After an hour we were tired of standing and went into the bar where we talked to the bartender. He was employed in his day job at Capital Center, a flight following agency. By 11:00 we were bushed, so we went to the transient quarters where we stayed overnight, and caught a flight to Binh Thuy the next morning.

I worked at the hangar the rest of that day, and early that evening Diamond came by the hootch to tell me we had to work on two of our gunships that were late for their missions. By the time that was finished it was 1:00 AM and it was back to the hootch for some much needed rest.

Looking back on our experience, I concluded we were either lucky or someone was looking out for us. The visits to Phuc Quoc Island, Hawaii, and now China Beach suggested we were living the *Life of Riley* (a TV comedy from the early 1950s.) But no matter where I went or what I did, it sure didn't feel that way because there was just no escape from the oppressive mental conditions. Being involved in war, a kill-or-be-killed proposition, saturated thought and I knew there would be no relief, we all knew there would be no relief, until we were back in The World.

**Remembering Gerry**

CPT Gerry Wick of Lockport, NY was tall, slim and slightly stoop- shouldered, as fine a person as there ever was. Gerry had a black brush cut and a friendly, even temperament, including a ready smile that came easily and often. We were flight school classmates at Fort Wolters, but Gerry went on to Fort Rucker for the second half of flight school and I went to Savannah. Following flight school he transitioned into Chinooks and went to AMOC, finally arriving in Vietnam in mid-December 1970.

Gerry was assigned to A Company (with the radio call sign Pachyderms) of the 159$^{th}$ Assault Support Helicopter Battalion (Chinooks), which was assigned to the 101$^{st}$ Airmobile Division in Thua Tien Province, just north of Danang. On 15 February, two days after we returned from China Beach, Gerry and four other crew members were killed when his Chinook crashed while performing a maintenance test flight in bad weather. They had apparently gone into IFR conditions and crashed nose low, almost inverted. The Hook went down very close to the coast of the South China Sea, north of Quang Tri.

I didn't find out about Gerry's death until almost a month later when my wife sent me the news which she had read in *Stars and Stripes* newspaper. We were both very sad about losing Gerry, and felt terrible for his wife. He was a fine officer, and a fine man.

**The Knife**

Our second shift maintenance mechanics were supervised by a sergeant who was overweight, with a round stomach and face to match. I didn't know much about him, but his southern drawl said his origin was south of the Mason-Dixon Line. He seemed to get along reasonably well with the mechanics.

Around 10 PM one night as I was waiting for a Huey to come out of PE for test flight, he came up the stairs to the maintenance office and said that one of the mechanics was hopped-up on drugs and was disrupting the work crew. He didn't tell me he was also waving a knife around.

A minute later the mechanic appeared at the top of the stairs with no shirt on and a 6" knife in his hand. He was a man of slender build, about 5'8" tall, but with the knife in his hand he looked much taller. I was sitting on a desk with my feet dangling off the side as the mechanic approached the buck sergeant waving his knife. It was obvious he was doped up on something by the way he walked and talked, and in slurred speech he said something to the sergeant about leaving him alone.

The sergeant froze in his tracks. I said, "You'd better put that knife away and go back downstairs," but he just looked at me with a scowl as he swayed back and forth. Looking again at the sergeant, he said something incomprehensible. The sergeant repeated what I had said, and a moment later the knife-wielding mechanic turned and walked down the stairs. We both breathed a sigh of relief, and I told him to call the 1st sergeant and have him call the MPs, or come himself and get the guy out of the hangar.

The 1st sergeant arrived a few minutes later and the knife-wielding mechanic was eventually led away. The next day I asked if charges had been brought, but was told that the company couldn't spare to lose another mechanic from a crew that was already short of personnel and that everyone was just going to forget the incident. It was just one more reason to get the hell out of the place, as far as I was concerned.

--- Ω ---

On 13 February, orders were issued awarding me another bronze star, this time for "meritorious service in connection with military operations against a hostile force." Because there were five recipients listed on the order and the timeframe for each award covered a long time period, I assumed this was a customary recognition for pilots who were close to DEROS. This bronze star was nice to get, but it had little meaning for me at this point in the game and amounted to little more than a nice gesture.

--- Ω ---

A few days before the orders were issued for the bronze star, battalion orders were issued directing Al to Ft. Bragg, NC on 7 April. The same orders directed me to leave Vietnam for Ft. Riley, KS on 1 May. These orders were the first tangible evidence that there would be life back in the States after the war. Holding them in my hands and knowing I still had almost three months of combat to get through produced an odd mixture of feelings. I wondered if it was too soon to begin hoping to go home in one piece.

--- Ω ---

On 14 February, Valentine's Day, I flew with an instructor pilot who declared me qualified to be an aircraft commander (AC), and the next day I recorded my first flight hours as an AC. From that point on I was fully in charge of, and responsible for, whatever my crew and my aircraft did. As President Harry Truman might have said, "The buck stops here."

**Wingnut 6**

Soon after I became an AC, Diamond Jim departed for home on a 14 day leave which implied he would be back in two weeks.

But I knew better. He had been warning me for months that this would happen, and now it was fact. I don't recall any goodbyes being said, but I do recall feeling a definite loss. In only a few months we had become partners who trusted each other's judgment and looked out for each other. I would miss his energy, enthusiasm and aggressiveness, but not the chip-on-the-shoulder attitude he always had. This expected development left me, Hardy and Woodley to handle everything by ourselves.

By 22 February everyone knew that CPT Diamond was not returning and I officially became the service platoon commander. As the official Wingnut 6 and still a lieutenant, I was placed in a position usually reserved for an experienced captain with formal maintenance training. The maintenance of 35 Hueys and the welfare of the 40-odd men in the platoon who maintained them was now my responsibility. And the battalion commander still expected us to provide maintenance support in the field whenever needed. This demanding environment of nearly non-stop work with brief snatches of food and rest looked like it would continue indefinitely, but with one less pilot with which to do it.

As I moved into Diamond's role, Hardy, having no combat experience and precious little flying experience of any kind, moved into mine. As a result, we flew together on maintenance recoveries, and during these missions he would plead with me to train him as fast as I could so he would know what to do and how to do it when I left in May. I understood his anxiety, but offered little sympathy. As the rest of us had, Hardy would have to develop a mentality of fearlessness if he was to survive the war. And that could not be given to him by training; he could only gain it through experience.

## Exasperation

Wingnut 6 was a role I secretly relished, but with the demanding schedule and responsibilities I lost weight rapidly trying to keep up. Woodley was still complaining about the pace of the work and nagging me about getting more help, but there was nothing I could do about our shortage of pilots. I was relying heavily on the outstanding work of the platoon sergeant to keep up with the non-stop rotation of Hueys in and out of maintenance, and the days blended together seamlessly into an endless string of test flights, maintenance recoveries and administrative demands. Hornblower and Smiley were magnificently reliable and relieved a good deal of the pressure generated by the situation, but there was almost no time to see Al, Tom or Mike, or for much recreation of any kind.

We were "grinding it out," but we were not the only ones. The enlisted men assigned to the platoon had as tough a row to hoe as we did. Normally the platoon had about 40 mechanics who worked 12 hour shifts, seven days a week. But due to attrition from men going home, and with almost no one to replace them, we were down to 21. On top of that they each had guard duty every third night with the following day off, leaving us with a skeleton crew on each shift, not even enough to keep up with the workload.

The missions didn't slow down, however. Day after day there were more Hueys waiting to go into PE. Some of them needed very little work and were back out on the flight line in short order. Others came into PE in such bad shape from age that they had to be shipped out for overhaul, and we would lose them altogether.

Every day was a rat race to see if we could keep enough Hueys in the air to fly the assigned missions. The mechanics would no sooner get one out the hangar door than two more would take its place. In one case a Huey left the hangar and flew only two missions before the pilot flew it into the ground at night. No one was hurt, but the Huey burned and was a total loss. With this combination of factors working against us, the 1st Platoon eventually woke up one day and had only one Huey in flyable condition. It was not a good situation.

My platoon sergeant and I juggled the work schedule as best we could, but there was no real possibility of it getting better until we got more people. Every day MAJ Coleman was on me about the problem, and every day I would explain the situation to him again: it was a lack of man power, and if he would get us more mechanics we could resolve the delays. But all he would say was, "Someone's not doing his job." I began to think he was as hard to talk with as Simon had been, and looked forward more and more to recovery missions that took me away from the problem. When I took off on a recovery with Hardy, these problems went away and I could get away from Coleman for a while.

Coleman's ass chewing was never appreciated, but I knew he had a job to do and was probably as frustrated about the lack of personnel as I was. But what really bothered me was the degree to which we were becoming vulnerable. The men were working too many hours and too many days without breaks, and the platoon sergeant and I worried that exhaustion would result in careless work or mistakes that could have disastrous consequences. The situation seemed to get worse every day.

For me there was only one way to handle the helpless feeling that came with this situation, and that was to lend a hand. I spent more and more time every day learning how the mechanics did their jobs and what I could do to save them time. I wasn't qualified to handle a torque wrench, but I was able to safety wire a nut, or replace a dead battery. This hands-on stuff was what I'd learned from my family and it provided some sense of accomplishment, certainly more than arguing over things I couldn't do anything about. But it didn't relieve my concerns about potential mistakes. I had to hope that if they occurred, they would be innocuous.

**CPT Les Bergman**

Les Bergman was a tall, blond haired, handsome captain from Wisconsin who had been assigned to the 191st a couple of months earlier from the 121st in Soc Trang. He was detail-oriented and always dressed like he had just stepped out of a tailor's shop, a perfect fit for the Green Delta Platoon that flew the VIPs. Les was friendly, liked to laugh, and enjoyed a good drink. Most of the time he was agreeable and reliable. But, if provoked he could display a hair-trigger temper which was often accompanied by aggressive, in-your-face behavior that intimidated many.

One morning about 4:30 AM, while trying to catch up on some badly-needed rest, I was awakened by someone pounding on the door of the hootch and yelling at me to get up. It was Bergman, and his hair-trigger clearly had been pulled. He had been doing a preflight inspection of his Huey for that day's missions and found a mechanic's wrench inside the tail rotor shaft compartment. It had evidently been left there by one of the

Wingnut mechanics who had been working on the Huey. With the wrench in his hand he began to rant about how irresponsible it was that it had been left inside the compartment and how the person responsible for it needed to be punished. Moreover, he was angry at me for letting this happen and wanted to know what I was going to do about it.

With a full head of steam, he went on this way for a bit and I couldn't get a word in the conversation, until he paused to take a breath. Seeing my opening I responded that this situation was exactly why we, as pilots, were supposed to do a preflight inspection, and that tired mechanics on understaffed crews were bound to make mistakes in spite of the best intentions. I also pointed out that we should all be hopeful that this would be the worst error we would encounter during these difficult times.

Apparently I said the right thing because Les immediately calmed down and a smile returned to his face as he somewhat reluctantly agreed with me. He left and went back to the flight line and later in the day I briefed the platoon sergeant about the incident. I could see he was troubled by it and then voiced his frustration about the stress on the maintenance crews from our short-handed situation. But while we agreed about the situation, there was still nothing we could do.

As we had been before the incident, Bergman and I went on to be friends and were later stationed at Ft. Riley together where we occasionally played racquetball (he always won). But questions haunted me: was this the worst mistake we would see? Was this the beginning of a trend? What could go wrong in the future? It was nerve wracking.

## Angel in Can Tho

The end of February was approaching and I would soon have less than two months left on my tour of duty, yet I hardly felt ready to return to the States. The daily routine, as difficult as it was, had become normal, and I didn't like to think about the adjustments that would be required when I got home. The unrelenting urgency of combat, and the life or death consequences of daily decisions, had almost become necessity, producing a nonstop adrenalin rush, a constant high. I was hooked, and realized nothing would be able to replace this environment.

Coupled with this was my recent R&R experience that pointed toward a hard adjustment upon returning home. And this fact, along with a growing respect, and even love, for the Vietnamese people, had turned my internal emotional landscape into a jumble of conflicting feelings. How could I love people I could not trust and who were out to kill me? Why should I care for a people I had been sent to kill? Why did I feel understood here, but possibly not in my own homeland?

But more than anything else I was lonely, already anticipating the loss of people who had become mainstays of daily survival: Al, Tom, Mike, Gregg, the guys in the maintenance crews, the other platoon leaders. Against all desire, against all will power, time was going to break up this team I had grown to rely on for support and survival, and every day was another day closer to this eventuality.

Late in February, an unprecedented trip to downtown Can Tho was arranged for the commissioned officers in the company. The plan was to take a company truck into the city, have dinner at a hotel, and return to the airfield in time to have a good night's sleep before missions started the next day. It was designed to be a safe evening out on the town for officers who wanted to experience the local color, and we were up for it.

About 4 PM on the appointed day, a group of twenty officers climbed into the back of a truck that drove out the main gate and turned right on to the road that led to downtown Can Tho. In ten minutes the truck was in front of the hotel where we jumped out and went into the dining room for drinks and hors d'oeuvres.

We sat along tables that had been pulled together to seat all of us in two long rows. Large, fresh prawns and other appetizers were passed around, along with drinks of all kinds: beer, rice wine, mixed drinks, nearly everything we could imagine. Several Vietnamese waitresses served us and I noticed one of them was a round-faced girl named Tam (*tahm*) who was a waitress at the O Club at the airfield. I mentioned this to Bob, who was sitting next to me, and he said he had known that she worked here as well as the O Club, and that she actually lived in the hotel. I didn't have any idea how he knew these details, and then our attention turned to ordering dinner from the Vietnamese menu (with a lot of help from the restaurant staff!)

The multi-course dinner was excellent and afterward we had more drinks, and then a few more, until we all were well-oiled. By that time the sun had set, and the beautiful, cloudless, blue sky had disappeared into a murky, oriental darkness. Across the street a few dim lights could be seen as we peered through the front windows that lined the front of the hotel. The darkness was intimidating as the street crowds from the day had evaporated along with the daylight, and as night had fallen an anxious uncertainty had crept into the group that was given away by the nervous glances of the officers. No one said anything about it, but the looks on their faces gave away their growing discomfort.

All at once the company truck pulled up to the front of the hotel in a cloud of diesel fumes, its bright headlights temporarily dispersing the darkness. As we began shuffling toward the door, Bob grabbed me, saying something about not being ready to go back and wanting me to stay there with him. I didn't understand what he was doing as he tugged on my arm, encouraging me toward a set of stairs leading to the second floor.

With each tug his voice became more emphatic, and I didn't know what to do. I didn't want to desert him if he truly needed me to stay, but I was worried about being away from the airfield at night, and I figured that if we didn't take the truck back now there would be no way to return until daylight the next morning. On top of that, I was concerned that something unexpected might happen and I would be needed for a maintenance recovery, or a test flight, or God knew what else.

I mentioned this to Bob as the rest of the officers filed out the front door and got on the truck, and I could see the choice between staying and leaving evaporating quickly! Bob continued to tug on my arm, but the alcohol I'd consumed, the uncertainty of the situation, and Bob's urgent coaxing had me confused. All I could think of was everything my imagination told me could go wrong if I stayed, and I heard a voice in my head saying things like, "Are you crazy? You could get court-martialed for not being on duty tonight"; and, "Enemy VC live in this hotel and will slit your throat if you stay." However, another voice was saying, “Look, Bob wouldn’t lead you into harm’s way, so what’s the big deal?” My head was swimming.

As he continued to tug my arm and coax me up the stairs, I watched the last of our group climb on the truck. Tam passed by us on the stairs with a coy smile on her face, and an attractive well-built Vietnamese girl on her arm. But being full of uncertainty, torn between staying and going back, I didn’t realize what was actually going on.

My anxiety and confusion peaked when the truck drove away to return to Can Tho Airfield and I realized I had passed the point of no return. I was committed to remaining at the hotel, and whatever that might bring. Turning from the hotel lobby, I followed Bob up the stairs, resigned to whatever fate the future held.

We climbed the stairs until we reached the third floor, and then turned down a dimly lit hallway that reminded me of the old movies I used to see about the Orient. Halfway down the hall we turned into an apartment, just as dimly lit as the hallway was, but with candles and incense burning around the room. I followed Bob into the room and Tam closed the door behind us.

In the dim candlelight I could see a low table with grass mats around it in the center of a room that was about thirty feet square. At the rear of the room was a small kitchen divided from the living area by chains of glass beads dangling in a doorway, beads that sparkled and danced in the candle light. Off to the left and right were two beds, surrounded by mosquito netting and covered with light blankets with beautiful oriental designs. There were several chairs around the room, and a bamboo couch next to the low table in the center. Bob was obviously comfortable with the apartment as he made his way to the couch and sat down next to Tam.

In a chair against the right wall of the apartment was the unknown girl who had passed us on the stairs. As Bob and Tam sat down she stood up, making a slight bow in my direction, and Tam introduced her to me as Thua (*too-ah*), her roommate. She appeared to be about twenty years old and was wearing a beautiful, floor length Ao Dai made of deep red silk with large white flowers that resembled peony blossoms on it.

I didn't have a very good look at her on the stairs, but even in the dim light of the apartment I could see she had a beautiful figure that could rival any of the girls we had seen in the floor shows at the O Club. Although I would not say she was beautiful, she had a very pretty, finely-featured face, and beautiful long arms that stood out porcelain white against the dark red of her dress. I was entranced by her, like a schoolboy seeing his prom date for the first time.

I glanced at Bob and Tam who were smiling at me from the bamboo couch, and finally put two and two together: this was a date (!), arranged for me by Bob and Tam. And with this realization came an inexplicable mixture of excitement and a new type of anxiety, suspecting I might soon have to decide whether to cross a line that I had struggled not to cross for almost ten months.

As my emotions bounced between confusion and exhilaration there was no room for reason, and what little self-control I had was evaporating quickly. My facial expression must have shown this internal struggle as I could see Thua responding to my feelings. The look on her face, a tenderly sympathetic expression of understanding, was almost angelic. In an instant all self-control was gone, replaced by a desperate desire that powerfully engulfed both of us.

| *Temptation Eyes* - Grass Roots |
|---|

For several hours she held and caressed me as a mother would her baby. She was gentle and accepting, calm and warmly encouraging even though she spoke little English and I even less Vietnamese. I was completely lost in her presence, my consciousness transported to a place of safety and refuge from the constant press of war. In her world I had escaped both the war and the anxiety of returning home to a place I knew would no longer understand me.

I awoke early the next morning, confused and disoriented. Thua was asleep next to me as were Tam and Bob in the other bed, and the old anxiousness was returning. I felt compelled to get back to the airfield and began to worry that Guardian 6 would be calling for Wingnut 6 and I wouldn't be there. I sat on the edge of the bed and quietly dressed, and as I put on my boots and laced them, she awoke and put her hand on my back. It was warm and soft (and inviting.) But I knew I had to leave soon or there could be unacceptable consequences. I hugged her and encouraged her to go back to sleep, trying to explain why I had to go. As I did, Bob awoke and started to dress.

It was already light outside and I knew the Boomerangs would be in the air by now, and it wouldn't be long before they might need my help. I didn't know what to do to express my gratitude to Thua, and the more I tried to verbally, the more tongue-tied I got. Finally, concluding that the words were not forthcoming, I did the only thing I could think of to express my feelings and left all the money I had in my pocket on the night table next to her bed. It was a crass substitute that would never express the gratitude I felt.

That one night with Thua was a profound experience, and today I wonder how she and Tam and the other lovely Vietnamese people I knew fared after we deserted them and their country. To me, Thua was an angel who offered love and mercy when I most needed it. And I have been, and always will be, grateful to her, and for her.

I can't remember how we got back to the airfield that morning, but when we arrived I was glad to find that I had not been missed, nor had my presence been needed for any emergencies. The only consequence of being out on the town was sweet memories of a graceful angel in Can Tho.

**Burning Vulture**

South of the U Minh Forest the battle for control of the southern Delta continued. There had been no let up from the daily assaults into VC controlled territory, nor in the damages being sustained since the operations had started several months before. If anything the fighting was more intense as time went by and the VC were forced deeper and deeper into the swamp land that made up the lower peninsula of South Vietnam.

At the end of February, we were called out on standby in case maintenance recoveries were required for the day's operations south of Ca Mau. We were making circles in the sky above the primary landing zone as the Vulture flight of the 162nd was making its exit from a hot LZ, when all at once the tail boom of one of the Vulture's Hueys caught fire about 200 feet into their climb. The Huey leveled off for a few seconds and then fell almost straight down, hitting the ground as flames completely engulfed the aircraft. It seemed impossible that anyone could have survived the concussion and fire.

About a month later, I was appointed to a promotion review board with two other officers, one of whom was a pilot from the 162nd. As we were getting organized to interview candidates for promotion, I began a casual conversation with the other pilot, mentioning that I had seen the chopper and crew destroyed just a few weeks before. He told me that he had been the AC on that Huey, and that no one had died in the crash.

I was most surprised and asked him how that could be possible. He said that when the Huey was hit by ground fire, the fuel cell had been ruptured which then caused the tail boom fire. As the fire grew, it affected the controls and the chopper started to lose altitude. When the Huey neared the ground he and the copilot pulled pitch, and also pulled back on the cyclic as far as it would go, slowing the Huey just enough that it softened the crash, and the crew was able to jump out nearly unharmed. I was amazed at his story, which was another of the many examples of unexpected outcomes that came from flying helicopters in combat.

**Hangar Queen**

Every Huey had a serial number that was stenciled in black letters on its tail. Each number had seven digits with the first two at the left divided from the last five on the right by a dash. In most cases each aircraft was referred to by the last three numbers (e.g. 68-16104 was known as 104). One of the older Hueys in our company was 109, a Huey that was in the hanger for repairs so often it was referred to as a hanger queen, and it seemed like Murphy's Law applied specifically to this Huey: If something could go wrong with it, it would!

On Saturday, 27 February, good old 109 was in the hangar again, sitting on jacks while one of the landing skids was being worked on. With no warning our "queen" slipped off her jacks and the fuselage hit the concrete floor with a bone-jarring crunch. I was in the hangar at the time, and the sound sent chills up my spine, knowing immediately what the sound was. No one had been hurt, but 109 had sustained serious structural and sheet metal damage.

This accident had been caused by human error, no doubt due to mechanics being tired or in a hurry, or both. After hearing the news, MAJ Coleman arrived a few minutes later to see for himself what had happened. He looked it over, said nothing, then turned and got back into his jeep and drove off. I expected another tirade, but that didn't materialize, and after this incident he was a different person. Perhaps he had finally realized that what I was telling him about our personnel shortage situation really was affecting the quality of our work, and the safety of everyone. Maybe now we would get some more mechanics, I hoped.

# Chapter 11 - MARCH 1971

Two months to go.

| *Look What They've Done to My Song, Ma* - Melanie |
|---|

**Set Up**

Ever since he took over the 191st from Blake Simon, Ignatius Coleman showed he was different. He was not arrogantly distant as Sechrist had been, nor was he hopelessly over his head like Simon seemed to be. He was pushy, in a benign sort of way, and hated to be told he couldn't have what he wanted.

Every few days Coleman would have a meeting of his primary subordinates, a staff meeting so to speak. Present were the four platoon leaders, the operations officer and the supply officer, and these people came together to get direction from Coleman as well as give him feedback about how they perceived things on just about any subject. Diamond Jim had been in on all the meetings before he left (although he disliked staff work passionately, including these meetings), and when he left I took his place on Coleman's company staff.

In one of my first meetings, the issue of President Nixon's announcement that the U.S. was getting out of the war and turning it over to the Vietnamese came up. Coleman acknowledged the Commander-in-Chief's prerogative to handle the war as he saw fit, and emphatically stated that since we were leaving Vietnam he didn't want any of us, or any of the air crews in the company, taking unnecessary risks. He wanted everyone to go home, and no one to be killed on his watch. While these words were received well, I was lost as to what actually constituted an unnecessary risk. It seemed to me that the entire war was an unnecessary risk, and I wondered just how the South Vietnamese people could possibly succeed against the North Vietnamese without us.

One day early in March, Hardy and I returned from being out on maintenance recovery duties for the battalion, and found there were numerous loose ends that needed to be tied up. After checking the status of maintenance work at the hangar, I called the company clerk and told him to tell MAJ Coleman that I would be late for the staff meeting, then went to the flight line to work on a couple of Hueys. When these daily chores were done, I went to Coleman's hootch where his meeting was in progress.

As I walked toward the hootch, conversation and laughter could be heard coming from inside, a good sign after a hot day out in the AO and with a lot of work on our own birds still to be done before the day was over. As I walked in, the conversation abruptly ended and I found a place to sit. Coleman made a sarcastic comment that it was nice of me to show up, and couldn't I read my watch well enough to be on time for the meeting? He went on to say that everyone else had been on time, and so should I.

I mentioned that the company clerk had been called to tell him I would be late, but it didn't seem to matter. I mentioned the maintenance recoveries we had been on all day, but he still glowered at me. And finally I brought up the Hueys we were working on after getting back to CTA, the ones that were needed for tomorrow's missions. But I was getting no sympathy.

Then, in front of the group, he said I needed to learn how to do less myself and get my people to do more. And if I couldn't do that, it reflected poorly on my leadership skills.

The other guys in the room, with slight smiles of satisfaction on their faces, sat quietly and watched me being fried. It felt like Coleman had shoved a knife in my back, and I was totally disgusted and angry with this treatment. But then he started to twist the knife. Going on with his criticism in spite of my objections, he wound up saying that if I was so busy I couldn't get to his important meetings on time, I should just get out and go back to the flight line and take care of business there.

I took this comment as an invitation, and picked up my cap and left. I was steaming when I reached the flight line, and stayed that way for the next couple of hours until we had caught up with what needed to be done.

Needing a beer to cool off I walked to the Boomer Lounge and found only a couple of people when I walked in, and one of them was Mitch Gregg, the company operations officer. As soon as I walked in the door he started laughing, so I asked him what was so humorous. He replied, "You are." I didn't know what he was getting at, and asked him what he meant.

Gregg went on to tell me that Coleman had set up the incident in the staff meeting before I got there, as a prank. Before I arrived, he told the others that I was the best maintenance officer he had ever seen and that I worked my backside off to see to it that we never had to turn down missions because the company's aircraft were always available. Gregg went on to say that, not only had Coleman been disgustingly profuse with his compliments about my work, he also said he knew I was short tempered and it was easy to yank my chain. Like a fool I had played right into his trap and Gregg and the others thought it was hilarious.

I didn't know what to think about this story. I didn't like being the butt of anyone's joke, but was pleased that Coleman thought highly of my work and leadership. No one else mentioned the incident, which I took as a confirmation that I was doing the right things and making the right choices, at least as far as the CO was concerned. I hoped there would be no more set ups.

## 3rd M.A.S.H.

One of the most important people in the company was our aviation maintenance technician, or tech rep, as we called him. The tech rep was a very highly trained senior warrant officer who knew more about the Huey than all the rest of us in the company combined. And he had the authority to match his rank: if he said a Huey was unfit to fly, it didn't fly, period.

Our tech rep was a very fine guy from Massachusetts who had many years in the service and the respect of everyone in the maintenance platoon. On Wednesday, 10 March, we were talking in the hangar office when all at once a deep, intense pain developed in my abdomen. Within seconds I passed out at the desk where I was sitting, and came to a few moments later with the tech rep asking if I was alright. I didn't know; nothing like that had happened before, but the pain was still there. Sincerely concerned, he suggested I go to the airfield dispensary to be examined.

As I left the hangar, the company Executive Officer, CPT Jerry Letterman, drove up in his jeep and I asked him for a ride to the dispensary. Walking in I explained my problem to the flight surgeon, who said he thought I needed more help than he could provide, and gave me a paper sending me to the 3rd Surg (the 3rd Mobile Army Surgical Hospital, or 3rd M.A.S.H.) at Binh Thuy AFB. CPT Letterman drove me to the hospital which was about 10 minutes by jeep, and dropped me at the place we had flown wounded GIs into many times in the past. I never expected to be there myself.

In the emergency room a doctor looked me over and said he didn't know what was wrong, so I was x-rayed and given an EKG, then given 100cc's of Demerol to stop the pain. The doctor was troubled by my symptoms, so he held me overnight for observation in the ambulatory ward. I was really uncomfortable being held there, knowing that Rip Hardy and Dex Woodley needed my help, and worried about what might happen without me. But the doctor was insistent that I remain, at least overnight.

The 3rd Surg hospital ward was like a prison, ruled over by Army nurses who were tough as nails. They gave orders the same way to everyone, regardless of rank, and they took no guff from anyone. If they delivered you a meal, you were expected to eat it, all of it, no excuses. And in the morning they came through with clean sheets that were tossed on each bed with directions to the patients to get out of bed and change the sheets. There was no coddling, and there was no question as to who was in charge. At first I didn't understand such tough treatment from females, but before long I saw its value in preventing patients from feeling sorry for themselves.

The following afternoon Gregg and Hoppes came in with my toiletries and some writing materials, and gave me some friendly harassment about goofing off. After a restless night with an IV stuck in my arm, I awoke, ate breakfast, changed my bedding and then waited for the doctor to come by and tell me what was going to happen next. I didn't feel poorly, except I still had a dull pain in my lower abdomen.

A new set of x-rays were taken, and when the doctor made his rounds he prodded my belly with his hand and said I had a kidney stone, which was not unusual because of the high heat and high minerals in the local water. That was all the information I got that day, which I spent much of worrying about the guys back at the 191st.

Early that same morning, Connie Leopard stopped in to talk for a few minutes and an hour later Coleman, Letterman and Gregg dropped in to check on my progress. About noon Mike Tschudi came by with a Get Well card and we talked for a few minutes. He said Tom and Jack Wyatt were planning to stop in that night. It was great to have visitors to break the monotony of the place, and to know that others were coming later. I was beginning to feel pampered, and in the early afternoon Dr. Miller* came by and told me I needed to be sent elsewhere for treatment. He also said something that completely surprised me: because I had more than 10 months completed on my tour, this ailment was enough to warrant sending me back to the States early. "You mean I'm going home?" I asked. "Yes," he said with conviction, "you are going home!"

This news, which was nearly beyond my comprehension, produced much confusion. Only 24 hours earlier I had been fully involved, physically and mentally, in the war. Now a physician was telling me I had been abruptly lifted out of what had become my daily torment, through no effort of my own, to be sent home to safety. In some cases it would be ahead of people like Al, who had arrived in Vietnam well ahead of me. This just didn't seem right, and all sorts of questions flooded my consciousness.

For one, how was I going to readapt to the world when I wasn't prepared for such an abrupt return? For another, how would I handle the possibility of being treated like other returning veterans who had been spit on and called "baby killers" by the very people who had either sent them, or allowed them to be sent to war? And what would happen to my friendships with a whole bunch of people I had grown to love and respect, my 191st comrades, as well as a growing list of Vietnamese natives I knew? There were no answers.

--- Ω ---

Besides the bossy nurses, one of the distinguishing features of 3rd Surg (and other Army hospitals as I would soon learn) was the ice cold temperature inside the place. After 10 months in country I had grown used to lots of heat and humidity, but in our ward there was neither. The air conditioning ran night and day and made it necessary to sleep under sheets and blankets to stay warm. This, by itself, was ample reason to want out of the place as quickly as possible.

Late that afternoon an infantry captain who had been on outpost duty was assigned to the bed next to me. We talked for a few minutes and I found out he was there because he had an attack of malaria that they were trying to get under control before his fevers killed him. In the middle of my second night at 3rd Surg, I awoke to find several nurses dashing to and from the captain's bed and throwing his blankets and sheets on the floor. They were frantic in their work, and I asked one of them what was going on. She said the captain had a temperature of 106 degrees from a malaria fever that had started during the night.

While I didn't completely understand the seriousness of the situation, I knew it was not good. As I watched them work an orderly wheeled a cart to the bed. On the cart was a small electrical refrigeration unit which they plugged into a wall outlet, and from the unit ran several hoses to what looked like an aluminum blanket. As the refrigeration unit kicked on, the nurses ceased sponging the captain with ice water, and within a couple of minutes his temperature had fallen back to something acceptable. It was an amazing display of expertise and technology and the next morning I was glad to see the captain showed no signs of ill effects.

--- Ω ---

After breakfast and a bed change, I found myself talking to Dr. Miller again. He said they could not treat the stone effectively at 3rd Surg, so he was sending me to Saigon ASAP. That day Jack Wyatt and Dave Mahala dropped in again, and later on Hoppes, Gregg and Letterman also came by. I told them the doctor was now saying I'd go home for sure, so Al said he would pack my personal stuff to mail back home.

Gregg, ever the joker, said he had got a large rock and painted "Barth's Stone" on it, and then put it in the urinal in the officer's latrine. Everyone had started urinating on it in hope of getting a kidney stone and going home too! There's nothing like Army humor.

## 24th Evac

On 15 March, Mike Tschudi called me to say goodbye. He was leaving for home 20 days early, courtesy of the Army. I would certainly miss him. He also said Tom was coming over that night to say farewell. In the afternoon Coleman, Hoppes and Gregg came over to say goodbye, and that evening Tom and LTC Williams, the battalion CO, came by as well. I was grateful for all the warm wishes these guys expressed, and felt terrible leaving them behind.

The next morning I was loaded onto a stretcher, then into a C-130 and flown to Bien Hoa AFB where an ambulance picked me up and drove me to the 24th Evacuation Hospital. I immediately went to the urology clinic, and after that was assigned to a bed. That afternoon I met with a tall, slim major doctor named Sorenson*, dressed in jungle fatigues and a stethoscope, who ordered the IV taken out of my arm. Then he dropped a bomb on me and said it was unlikely I would be going home because he would remove the stone if it didn't pass by itself. This was a real shock as Dr. Miller at Binh Thuy had said I would not be returning to flight status or CTA, but would definitely be going home.

After the interview I went to the small PX at the hospital to get some reading material to help dispel the monotony that was part of the experience of a combat hospital, and to cheer myself up after getting Sorenson's news. I picked up a copy of *The French Lieutenant's Woman*, a 467 page novel by John Fowles, and a best seller at the time. It was the only thing that looked interesting on the rack, and being a slow reader I figured this would keep me occupied for quite some time. It turned out I would need all 467 pages, and then some.

This situation was typical of the Army and its emphasis on process over efficiency, and the whole thing was becoming comical. Lying in bed waiting for a stone to move was like the joke about a 1st sergeant who, as punishment, had assigned a GI to watch the rocks lining the walk to the company headquarters and report if any of them moved on their own. It seemed ridiculous.

Days went by, and to pass the time (since I had trouble passing the stone) I took to exploring the hospital's various wards. These excursions were jolting and depressing, seeing the condition of GIs who were there for reasons much more serious than mine. There were missing arms, missing legs, and bandaged heads (sometimes with heads completely hidden by bandages.) There were GIs who had had sucking chest wounds and were fighting to stay alive.

But the absolute worst, bar none, were the burn victims who had to have frequent and painful changes of dressings to prevent infections. They were in the worst condition, so I thought, until I saw a Vietnamese man who had lost both arms and both legs, and was left with nothing but a torso. He could do nothing for himself except lie in his bed and moan, and every few hours a nurse would come by and turn him over. What would this man's life be like in the future? God help us, I thought to myself.

From a letter dated 20 March 1971, 7:00 AM:

> ***Good morning! If you are wondering what I am doing up already at this ungodly hour, so am I. They***

turn the lights on around here at 5:00 AM and if you are not in some state of awareness, or a close resemblance, all the corpsmen go by and kick your bed-- not out rightly. They do it very discretely as if they just happened to accidentally nudge it as they go by. Then if you are not awake by 6:00, they start talking very loudly. By 6:30 they start verbally attacking you and by 7:00 they bring out the tear gas and riot guns. They don't really do that. I just made it up. But they do insist that you get up before 7:00 and make your bed. If you go back to sleep after that, they don't care. It's usually quieter between 9 and 10 than nearly any other time of the day. So, it's always a good time to sleep.

Lately I haven't been able to sleep come 10:00 PM when they turn the lights out. Guess I'm all slept out. I just wander around the ward, eat some ice cream, smoke a cigarette, and watch other people's TVs. Sounds great doesn't it?

The guy in the next bed, also a lt., has the same problem that I do-- except he has quite a bit more pain than I ever had. He just got a shot of morphine and is going cross-eyed to sleep. Must be great stuff.

When I got here the only available bed was a cot. Not just any ordinary cot either. It's pre-Civil War. It slept like a hammock and moaned in its sleep. It was also giving me a backache. So, yesterday afternoon I moved to a new bed-- another cot, but this one has a bed board. Now it feels like I'm sleeping on a blanket in the back end of a GMC truck. The pajama bottoms I've got *were made for an extra portly so that when they are laced up there is a big bunched up knot in front the size of an egg. Every time I rolled over last night it felt like I was sleeping on a big marble. Army hospitals are so wonderful. Think I'll put in for a PCS to this place for the duration. Hah!*

The days turned into a week, but the stone was unmovable. Each day I would go to the toilet and sit, waiting for some sign of progress, but nothing developed.

From a letter dated 21 March 1971, 5:30 PM:

*The local Hippocrates was in this morning. Said he was looking for stones, unfortunately I had none to offer.*

Doc Sorenson came by on Monday, 22 March, and said that if the stone had not passed by Wednesday he would go in and get it. Two x-rays that day (the first one was messed up) confirmed the continuing presence of the stone, but I could have told them that since I could feel it in my bladder. The lieutenant in the bed next to me, an animal science major from North Dakota State, had the same problem I did. But that didn't stop him from jumping out of bed and crossing out "1LT" and writing in "CPT" on my bed card the first thing that morning. It was promotion day for me, and everyone who went by said, "Good morning, captain," with a nod and a smile. It felt great to no longer be a lieutenant.

The next day, the head nurse at the 24$^{th}$, a lady lieutenant colonel, came by and pinned a miniature set of captain's bars on my pajamas and congratulated me. I was surprised and flattered that she knew about it and took the time to see me. Later that morning Doc Sorenson came by and said he would definitely operate the next morning. With nothing else to do that day, I continued to walk the wards and read my books. At that point I had read a small collection of classics and best sellers including *The Great Gatsby*, *From Here to Eternity*, *The Andromeda Strain*, and *The French Lieutenant's Woman*, and was just starting *Portnoy's Complaint*. I felt rested and recharged and was ready to leave this unplanned, two-week R&R behind.

The next morning I was wheeled into surgery at 8:00 AM, and 90 minutes later was back in my bed with a very sore penis and a strong urge to urinate. Before I left the recovery room the nurse said only a portion of the stone had been removed, and they wanted to see if the rest of it would pass by itself. I went to the commode and attempted to pee, but couldn't. That was followed by more waiting! But later in the day the rest of the stone passed by itself and the pain that had been persistent for two weeks was finally gone.

While I was in surgery, the lieutenant next to me passed his stone and the following morning was released from the hospital. On his way out he stopped by to see how I was doing and to say goodbye. He was but one example of the really nice people that the hospital was chock full of, but I was more than ready to say goodbye to them.

I had called the company the night before and talked to Al, and told him I needed a ride back to CTA. He was pleased to hear this news, and matter-of-factly said he would come and pick me up. He must have been drinking with a bunch of others because there was a lot of background noise, and I heard MAJ Coleman tell Al to tell me he was going to kick my ass when I got back (typical Coleman humor). But there was one problem: all my gear had been packed so they couldn't bring me any of my personal clothing.

On, Friday, 26 March, I was released from the 24th Evac and Al and Jerry Letterman picked me up at the 24th's helipad. They handed over what clothing they had found for me, which consisted of plain fatigues, a baseball cap and a pair of size 11 combat boots that I couldn't even lace up because they were too small. No socks, no underwear, no nametag, and no rank markings of any kind. I was anonymous and uncomfortable.

After I had dressed, Al, Jerry and I flew to Hotel 3 near Tan Son Nhut to have lunch before heading back to CTA. I objected because of the uniform I had on. But they were unmoved, saying I shouldn't be concerned about it and if anyone asked it could be explained. All I had to do was show my dog tags for identification. I acquiesced in spite of serious discomfort from the boots, and after landing we found a Mexican restaurant where the food was considerably better than the hospital fare of the last two weeks.

By 4:00 PM we were back at CTA. I was glad the hospital ordeal was over and things were back to normal, but the images of wounded GIs that I had seen were still with me. I thanked my lucky stars that it was a kidney stone that did me in, and not hostile fire.

**Captains, We!**

I unpacked the personal stuff that Al had packed up for me and climbed into a clean flight suit, checked on the operations at the hangar, and then headed to the Boomer Lounge. When I got there Gregg was alone preparing for the monthly promotion party, where all the newly promoted officers for the last 30 days were going to be recognized. Al, Tom Yost, Mitch Gregg, John Thompson and I were all promoted in March, so we financed the party, which included a semi-formal pinning on of our new captain's bars. LTC Williams came in to celebrate too, and Al and I did the cooking while Gregg managed the serving of all the other food.

Going from first lieutenant, or trung uy (*chung wee*) in Vietnamese, to captain, or dai uy (*die wee*), was a major milestone that provided a large amount of satisfaction. A person with the rank of captain was eligible to command a line company, a basic maneuver unit in the Army, and the critical building block for larger units like battalions, brigades and regiments. And a company commander had responsibility for 200+ troops, both their welfare as well as their performance.

There were lots of poor and marginal company commanders, and a good one was worth his weight in gold. I wondered if I could measure up to that responsibility. This promotion opened a door to command positions I would be assigned back in the States, not once, but three times in the future.

**Fire!!!**

The promotion party was grand and lasted well past midnight, until we finally collapsed into bed at 1:30. An hour and a half later, we awoke to pounding on our hootch door by Jerry Letterman who was yelling something about a fire. We grabbed our clothes, stumbled out the door and, and beheld a raging inferno engulfing the three hootches across the walk from us.

Jerry was yelling at us to get away from the blaze as far as possible. We needed no direction or encouragement to do that, because the ammunition in the burning officer's hootches was exploding from the heat and no one wanted to be wounded by a round from a fire.

We moved away from the blaze and watched as the three hootches and two nearby trailers went up like dried kindling. The ammo popping off sounded like popcorn as we stood and watched the flames leap 20 feet into the air, wondering if the fire would jump the walk and take out our hootch too. But it didn't, and within an hour there was nothing left of the three blazing hootches but smoldering rubble. The fire had consumed everything leaving only ashes in its wake. We were lucky; no one had been hurt.

**CWO Mike DePaul***

Most of the time we were concerned about encounters with the enemy, but sometimes our greatest threat originated from our own doing, or from people who were on our side. This was the case on 17 March while I was still in the hospital, for a warrant officer named Michael J. Depaul from Seaside Park, NJ.

Mike, whose call sign was Darkhorse 32, had been in Vietnam for 26 months, since January 1969, and was assigned to C Troop of the 16th Cavalry in Vinh Long. He was not only qualified as an aircraft commander in Cobra gunships, he was extremely experienced and served as the troop's instructor pilot (IP).

On the 17th he was giving a check ride in a Cobra to qualify LT Robert C. Green* as an AC, doing traffic pattern work and autorotations at Can Tho Airfield. Less than an hour into the check ride there was a miscommunication between Depaul's aircraft, the Can Tho tower operator, and a CIA fixed wing plane that resulted in a mid-air collision between the two aircraft and the deaths of all the people on board. It was another pitiful tragedy.

This was a particularly sad event for Tom Yost who had flown with DePaul in Soc Trang and regarded Mike as a top notch pilot. And it was a shocking incident to occur so close to home, a grim reminder to everyone that even the most competent and most experienced pilots were still subject to human error.

--- Ω ---

From a letter dated Tuesday, 30 March 1971:

> ***Things have really been screwy around here since my return. I told you about the fire we had. They've been cleaning up the debris from that and will start building at the end of this week. The next day, which was Sunday, there was some kind of major power failure at the power plant and it won't be fixed till Friday. In the meantime they've set up some temporary deal and we now have one hour of electricity followed by one hour of none. It alternates like that all day. It's all rather confusing, especially at night because the fans come on and go off by themselves. When they're off it gets real hot and the mosquitoes bite the devil out of us....***
>
> ***Last night everyone was timing it so that we would get the maximum enjoyment out of each hours worth of power. Get the records all set, turn on the air conditioners, and wait for the juice. Then when the***

*power went off again, up with the candle (only one). During one out period we had a tremendous trivia game that lasted all during the hour. There were about 10 of us and some really good, and difficult, questions were raised.*

--- Ω ---

On Tuesday, 30 March, I took a routine check ride with the company IP, and when we were finished I went to the hanger to do some test flights. It was almost as if there had been no interruption to the routine that existed before my hospital experiences. But under the surface some things were definitely different.

Being in Hawaii for R&R and then coming back, followed by being told I was going home early and then being sent back again, felt like I was riding a yoyo. On top of that there were pilots coming and going in the company so quickly I couldn't even learn their names. More and more I knew fewer and fewer people, and the company itself was growing foreign to me. And all the time the maintenance recoveries continued, admittedly less so than back in December, but still a daily demand with no let up.

I felt increasingly unsettled, anxious and alone, and wondered how I'd cope after the anchor of Al's friendship was gone in another week. I wanted to find some kind of stability to hold me through the month I had left on my tour, and the only way that could be done seemed to be a change in approach to the circumstances. The experience with Thua, my angel in Can Tho, lingered at the back of my mind. It seemed ironic that it was a native woman who had shown me the compassion and tenderness that restored my balance and stability, and I missed that feeling.

A day later I was in the officers club with some of the pilots and noticed a new waitress who was cute and somewhat shy. I asked someone her name and was told it was Hai (*high*), the Vietnamese word for two, meaning that she was the second child born to her mother and father. I also learned she was having a hard time emotionally because she had been the girlfriend of a pilot in the 162nd who had been recently transferred to a unit up north.

I wondered what it must be like for her to lose an American boyfriend, but didn't think about it anymore than that. I figured she might be a gold digger, looking for an American to marry who would take her to the U.S. with him, something many young Vietnamese girls aspired to.

# Chapter 12 - APRIL 1971

## Mr. Lucky

The brilliant white lights from the POW camp just beyond the concertina wire on the north side of the airfield were too bright to look at and still maintain night vision. Looking below the glare of the lights to my right, no one could be seen inside the camp.

I was alone in the dark, walking back from the group headquarters on a small path that paralleled the POW camp fence. Up ahead a narrow stream with tall grass lining its sides cut across the path in this nearly deserted part of the airfield. A dim, yellow light bulb in an old fixture mounted on a telephone pole next to the stream glowed weakly, trying to light the pathway where an old Huey rotor blade had been laid across the stream as a bridge.

I started across the bridge listening to the hollow thump my boots made each time they hit the rotor blade. About halfway across, I heard a loud buzzing noise pass by my right ear that I recognized as the sound of a bullet. I automatically ducked thinking someone had me in his rifle sights, expecting there might be more bullets where that one came from. A couple of seconds went by as I listened carefully for the sound of the rifle shot that should follow the sound of the bullet, but there was none.

I concluded it must have been a stray bullet fired from a good distance away that had just happened to come near me. If I had been walking slightly to the right ... I might not be writing this.

Call me, "Mr. Lucky."

**More than Brothers**

It was Sunday, 4 April, nine days after returning from the 24th Evac, and life was back to normal. The debris from the "Great Can Tho Fire of 1971" was being cleaned up and electric power had been fully restored. I spent the day at the hangar working with Woodley and Hardy, and early that evening, after scheduling the Hueys for missions the next day I went to the Boomer Lounge where Al's DEROS party was in process.

Before I arrived, Al had chugged a bottle of Mateus wine in a record 32 seconds, and when I arrived he was sloshed. Sitting on a bench with an empty Budweiser box on his head, a dart board under his shirt, a collective stick in his left hand and a cyclic in his right, he was telling war stories to the crowd that had gathered in the lounge to celebrate his departure. There was no one funnier than Al when he had one too many, and that night he was hilarious.

Yet even in the midst of all the fun, and being happy for Al that he was going home the next day, I was depressed. Al and I had been practically as close as two guys could be for the past eleven months. We had shared trips to Phuc Quoc Island, Saigon, downtown Can Tho and China Beach; war stories of near-death experiences; somber holidays; frustrations about the war and some of the people we had met; living quarters; football discussions; new friends; jokes; popcorn poppers; and losses of comrades. He had come to see me in the hospital, and picked me up on my return. I knew that when he left, a large part of me would go with him, and we might never see each other again.

What had all our experiences made of us? Something much deeper than friends, or comrades, or compatriots. Brothers is the closest word, but even that doesn't express the way we had each other's back, or the depth of sharing our feelings and aspirations that had taken place across these many months. To this day I cannot find a word to adequately express the relationship we had. It was a unique product of our shared war experience.

The next morning Al was in his usual sunny disposition in spite of the wine. He quickly gathered up his meager belongings. Then we shook hands and he said, "Well, Diggie, it's been real." I nodded, unable to say much of anything, and he turned and was gone. I felt like my right arm had been cut off.

**Hai**

Al and I had been in the same unit together since the beginning of flight school in June 1969, and for the last 11 months we had shared one small room ... but many large experiences. I wished he didn't have to go, but I was glad he was getting home in one piece. And I was grateful we had been more than friends for so long.

Mike Tschudi was gone to Germany by then too. And Tom was out flying every day as the battalion commander's copilot, which gave him little time for himself, much less time for us to get together. I seemed to be swimming in confusion and loneliness when I saw Hai again, waiting tables at the officers club.

Sitting by myself I watched her work, studying her manner. She was slim, petite, no more than five feet tall, with long black hair and a modest demeanor. She spoke English very well and when she came to my table I felt compelled to ask if she would come and talk with me when she finished work. She agreed, smiling warmly and batting her eyes, unable to conceal her interest.

An hour later she finished work, came to the table and sat down quietly. She was young and very attractive, delicate in appearance like the petals of a fresh flower blossom. She had polished nails at the tips of long, slender fingers, and when she smiled her teeth sparkled, pure white like the sand at Phuc Quoc Island.

At first there was an extended moment of awkwardness as we searched for things to say. But after a few moments we overcame our mutual shyness and a conversation began. I asked about her departed boyfriend and she described the feelings she had for him. It was a dumb thing to do, because for the next few minutes I felt like I was competing for her attention with his ghost, who was essentially sitting at the table with us. I tried to change the subject by ordering a couple of drinks, but it didn't work and she continued talking. She spoke honestly, from the heart. She said she knew her boyfriend was not coming back; she said she needed to move on with her life; she said she didn't like the loneliness his absence created.

I was not prepared for this blunt revelation of her feelings, and thought how unfair it was to both of us that she should be grieving for a departed pilot, someone I didn't know, but nevertheless was growing weary of hearing about. Gradually the conversation moved away from the ghost pilot and back to the two real people at the table.

As we talked, I felt like Hai's open vulnerability was opening a door to understanding her and her people, and I could do little to resist the temptation to explore the path beyond that door. I realized it could lead to a deeper personal involvement, but by now I didn't care. After all the death and hostility of the previous eleven months, I had needs too and I hoped Hai's tenderness might meet them.

The word that had guided my decisions up to that point was should, as in I <u>should</u> not do this, or I <u>should</u> be responsible, or I <u>should</u> be faithful. But now there was an internal voice that said, "You need to do what you want, not what you think you should. By doing what you should, you've missed valuable experiences that go with being in this foreign land." At that moment it sank in that Hai and I were not only pushing open the door to understanding, but also the door to an unknown relationship. And it was refreshing, exciting. There was no telling where it might lead, but any concern about that was distant. I wanted to be with Hai as much as possible to end the ache of loneliness.

The next time we saw each other at the O Club, we made plans to visit her home in downtown Can Tho in spite of the grief she was still feeling. Late in the afternoon a day or two later, I went with her to a well-kept, white two-story building where her parents lived. The whole neighborhood was well maintained and looked prosperous compared to most of the Vietnamese homes I'd seen.

She introduced me to her parents who lived on the first floor of the house. They spoke almost no English, but were very gracious in welcoming me. Then we climbed the stairs to her second floor apartment which was clean, bright and cheerful. The walls were white and the floor was covered with small, multiple-colored ceramic tiles that gave the place a colorful glow as the late afternoon sun reflected off them.

I sat in a bamboo chair as she lit some incense and prepared tea, and for the next couple of hours we talked about a lot of things: what her father did for a living, what the neighborhood was like where she lived, who her friends were and what they were like, what she wanted to do with her life, what her former boyfriend was like. It was clear that her affection for him continued, but it seemed like she was honestly trying to move beyond those feelings.

I enjoyed being around her optimism and youthful energy, along with her remarkable innocence and vulnerability. And I liked hearing about her family and life in Can Tho. We talked about her life some more, then when she was ready she asked me many questions: Was I married? Yes; did I have any children? No; when was I going home? Soon. But no matter what answers I gave, her interest in me didn't wane, nor did mine in her.

It had grown dark while we talked, so Hai lit some candles, softening the harsh feel of the white walls of the apartment. Throughout the afternoon and early evening she was sweet and non-judgmental, accepting and patient. As we talked, our mutual attraction grew, and then culminated in an unexpected mixture of desire and guilt. In spite of Hai's tenderness, it became clear that this pathway to understanding the Vietnamese culture was fragile, and someone could easily get hurt. Yet I liked Hai immensely and found her warmth and humor an antidote, at least temporarily, to the depression that had been haunting me.

| *Help Me Make It through the Night* - Sammi Smith |
|---|

The closeness we experienced at Hai's was precious, but a few days later I found out that her former boyfriend was on his way back to Can Tho. With that news I realized I would be out of the picture after he arrived, and it was disappointing to think I would not be able to spend time with Hai again. But I was happy for her, who with this news had lost her grief and seemed to regain her balance and composure. Besides, I reasoned, there was no way to have had any significant relationship with her when I was married and scheduled to go home in less than a month.

The brief interlude with Hai was as significant as any experience I had during the war. The qualities of her character and the situation she was in evoked a number of different roles. I felt like her protector, lover, father, companion, and friend. Our short relationship harnessed a complex of emotions that were confusing and rewarding at the same time. To me, she was special.

After many years and much thought, a different understanding of this experience has developed. In a figurative sense Hai was willing to show me the door into her world, but I could not pass the threshold. And in a larger perspective the same idea seemed to apply to Vietnam and the Vietnamese culture: I could be in it, but not of it. My experience with her delivered the clear message that there were real barriers to entering her world that I could never overcome.

What had most attracted me to Hai was her pure, childlike innocence in an environment where man's incredible capacity for hate and violence had escaped the confines of civility. She was beauty on the battlefield, a safe haven from eleven months of continual death and destruction. My desire for her and for her world was a concrete manifestation of how my thinking about Vietnam had changed, how love and compassion was replacing fear and hate.

The simple truth is we were two hurting souls who needed what we could give each other: relief from loneliness and uncertainty in a difficult time and place.

**Short Timer**

After my brief and exhilarating experience with Hai, I began to wrestle with thoughts of going home and having to confront a public that was either ambivalent about what I had done for the last year, or hostile to it.

It was close to the middle of April with only a couple of weeks to go, but that changed none of the daily mission requirements as Hardy and I were in the air everyday with test flights and recoveries. It was good to see he was quickly moving up the learning curve.

One morning we found ourselves en route to some obscure location for a maintenance recovery, as I had done so many times before, when I received a radio call from Guardian 6. He said a chopper had been shot down in a hot LZ and he wanted us to go get it. With only a few days left on my tour, going into an unsecured situation and potentially taking VC fire had grown less attractive by the minute. So I keyed the radio switch and said, "Guardian 6 Alpha, come up victor, over," meaning I wanted Tom Yost, who I knew was flying as Williams' copilot, to talk to me on VHF radio.

A few seconds later I heard Tom say, "Wingnut 6, this is Guardian 6 Alpha on victor, over." I told him that I didn't want to get anywhere near a hot situation on the ground with only two weeks left on my tour. Almost instantly I heard Guardian 6 say, "That's OK, Wingnut 6, I understand. I'll find someone else to handle it." Williams was monitoring the VHF radio and heard what I said to Tom, and was graciously letting me off the hook. I replied, "Roger, Guardian 6," with my best military radio voice. But inside I was thinking, "Thank God and Rudy Williams both!" Hardy and I looked at each other for a moment. There was relief written all over his face, and probably mine too.

**Air Force Blues**

Al's departure left an empty bunk in my hootch, and one day I got a message from the company office that an Air Force sergeant had arrived at Can Tho and needed a place to bunk for a couple of days. My hootch was going to be the place. A little while later I opened the hootch door at the knock of a black-haired sergeant in fatigues who was holding a duffle bag. He had a pasty complexion, soft, white hands, and was about 75 pounds overweight.

We talked a few minutes before he began to unload his gruesome story. As he told it, he had been stationed at a base up north somewhere that had been attacked and overrun by the NVA, who were eventually repulsed by counterattack. At the outset he had taken to a bunker with some others, one of whom was an Air Force captain who was hit by a rocket-propelled grenade (RPG) in the chest while standing in the door of the bunker. This sergeant was standing directly behind the captain when he was hit.

Over the next couple of days I heard, over and over, how the RPG round had blown the captain apart and left all sorts of blood and gore all over the sergeant standing behind him. He described the captain's arms coming off and flying past him as his body disintegrated all over him and the walls of the bunker. I didn't know what to say, so I just listened quietly. No matter how many times he told me this story, he could not seem to get the images out of his thoughts. He was a basket case sitting on Al's bunk, chain smoking cigarettes and retelling this grim experience that was obviously burned into his consciousness.

Now fully recovered from my experience with Hai, my old, hard-nosed survival mindset had returned and I didn't want to get drawn into sympathetic feelings. I felt bad for this man and thought he was on the verge of losing his marbles, but I grew tired of his gruesome story. I felt I couldn't afford to be pulled into his insecurity, and rationalized my distance from him based on the cold idea that we were in a war, and what could you expect from it? I eventually concluded that flying into hot maintenance recoveries for Rudy Williams was easier than listening to this sergeant's story again and again. So I spent as much time away from the hootch as possible. It was a relief when I returned one evening to find he was gone.

He was a different type of casualty, and I hoped he could find some help, but I certainly didn't know how to deal with the mental wounds he carried.

**The Dregs I**

War is pursued to damage or destroy an enemy. But, there is no way to contain or manage the collateral damage it causes as a consequence. One type of collateral damage is the corruption of the moral fabric of the warrior himself, and by extension his entire society. One did not have to look far to find examples of corrupt behavior among the U.S. troops in Vietnam, even among those of us who had been appointed officers and gentlemen.

For example, the same evening that Al's DEROS party was held, Jerry Letterman bit the head off a live chicken. That's right, with his teeth.

Now, this was not the friendly and gentle Jerry Letterman I knew, the man who had taken me to the hospital and then visited several times while I was there. This base, animalistic behavior was a product of too much alcohol and too little restraint and good judgment. It was a ghastly demonstration of sub-human behavior, a display of the corruption that can result from putting people into combat where they must choose between killing or being killed.

To me this act was moral idiocy, the bottom of the barrel, the dregs. It was disgustingly disappointing.

**The Dregs II**

Ignatius Coleman had shown himself to be a solid soldier and commander. He was predictable and mostly by-the-book, so when he was goaded into doing the same thing Letterman had done, and agreed to it, it was clearly out of character. I found it difficult to believe that he would actually follow through.

Word of the challenge got around quickly and it wasn't long before bets were being placed. The odds were even that he would actually do what he said he would, and I was inclined to believe he would not go through with it. The next evening a group of officers assembled to see Coleman and a scrawny chicken square off, and after several minutes of suspense, he grabbed the chicken by the head and body and bit clear through the chicken's neck. The onlookers howled like a pack of blood thirsty wolves.

I had put my hope in Coleman having enough good judgment to treat this as a way of baiting the crowd, and expected him to decline the act with a flourish. Having actually done it, he too had dipped into sub-humanity, corrupted to the same degree that Letterman was.

I took a cold shower and tried to wash the whole experience away, but it hung at the back of my mind like the odor of garbage on the streets of Can Tho hung in my nostrils. It was disgusting, appalling, barbaric. Who and what had we become?

**Goober Butter**

With more and more U.S. helicopter units turning over their operations to Vietnamese units, we began to see experienced pilots who had time left on their tours reassigned to our company. Being preoccupied with my platoon's work and spending a lot of time away from the airfield prevented me from spending much time with, and getting to know, the new pilots who were coming in frequently.

One of the recently transferred pilots was a tall, quiet captain with a big bushy mustache who I'll call Slim. I had seen him in the Boomerang Lounge a couple of times, but we had never talked. One evening Slim was in the Lounge with a group of pilots who seemed determined to consume all the alcohol at the airfield, and having a great time doing it. Eventually the crowd thinned out and Slim lay down on the couch and went to sleep. That was a mistake, and a cue for a practical joke.

Captain John Thompson had been in the company for quite a few months and was assigned to the Bounty Hunters as a gun driver. He was a likeable guy and a good pilot, but he had a mischievous streak that seemed insatiable. Seeing Slim in such a vulnerable position was more than he could resist, so he left and came back a minute later with a jar of peanut butter which he carefully spread on Slim's oversized mustache. He found a feather from somewhere and began to tickle Slim's nose with it. Of course Slim didn't wake up because of the alcoholic stupor he was in, but he did reach up with his hand in response to the tickling and smeared the peanut butter all over his face. Such great success only encouraged Thompson to spread a little more peanut butter and tickle a little more, and before he was done Slim had peanut butter all over his face, in his hair and in his ears. The onlookers were quietly, but highly amused.

The next morning Thompson came to see me and asked me not to say anything about what had happened the night before. He was afraid of retribution as Slim was in a rage and wanted to know who had done it. But Thompson was safe; no one wanted to spoil the outcome of a harmless and successful practical joke.

**Last Mission**

The 191st had a policy that pilots within seven days of leaving the country were not required to fly. Therefore my last combat mission was on 23 March when I flew two hours in a Huey, probably on a maintenance mission of some kind, but I don't remember the details.

Afterward there was a farewell party for several of us who were leaving in the next week or two. At the party, each of us was asked to make a toast and give a parting comment to the rest of the pilots who were there. When it was my turn I said something about the company not being the same as the old days when I had first been assigned to it last May. I did not articulate this very well, and the group took what I said the wrong way. Several pilots booed me for what I said and I felt like I had let them down. I wished I had taken more time to prepare something appropriate, words that would express the admiration and respect I had for them.

On the 26th, I turned in all my company gear to the supply office and said my farewells to Hardy, Woodley, the XO, and all the pilots and other guys in the maintenance platoon, and hopped a Huey bound for Long Binh. With me were Coleman, Tom Yost and Les Bergman, all of whom were DEROSing at the same time. As the chopper climbed away I looked back at the place that had been a lot more than my home away from home for almost a year. There were many memories there, good and bad, that I knew I would never forget.

Arriving at the 90th Replacement Battalion, we were assigned to a BOQ until a freedom bird would take us back to the States. Our nightmare was almost over.

# Chapter 13 - MAY 1971

## Freedom Bird

On May Day, Tom and I, and several hundred others, climbed aboard a World Airways jet and took off for the States. The atmosphere on board was happy but subdued, and when the wheels of the plane left the ground there was a short, muted cheer. The fear that had enveloped us on the flight to the war a year ago was gone. Combat had worn us out mentally, emotionally and physically, and the inexperienced young men from that day had become cynical, suspicious, wiser.

After a long, quiet flight, the plane landed in Anchorage, Alaska to refuel and we were able to stretch our legs for a few minutes in the terminal. Even in May, Anchorage was still buried in snow and I remember looking out the windows at it with deeper appreciation for the contrast it was to the deep greens and browns of Can Tho and the Mekong Delta. I was going back to a place where the people were used to this picture, but shared little of the understanding of the tropics I had gained in Vietnam.

We re-boarded and left Anchorage, touching down at Travis Air Force Base, California a few hours later. Tom and I splashed a little cold water in our faces to freshen up and then caught a bus from Travis to the San Francisco Airport where we got tickets for Columbus, Ohio on the first flight out. To that point we had encountered no direct public hostility, just a cold indifference from the people around us.

Soon we were aboard a flight to Ohio for what I thought was the last leg of our Vietnam journey. I was officially on leave for thirty days with no Army duty to bother with until I was scheduled to report to Fort Riley, Kansas on 4 June.

This felt like freedom; there was room to breathe again.

**More OERs**

When Jim Diamond left the 191st, he had no time to complete an efficiency report for me as his assistant service platoon commander (Wingnut 5). His report, which covered the time we worked together from September to February, actually arrived after I got to Fort Riley.

But it was worth waiting for. Both Diamond and Coleman rated my performance and wrote flattering comments that addressed my leadership qualities and organizational abilities, and recommended promotion and training ahead of my contemporaries. Any regular army (RA) officer would have been happy to see these comments on a war time OER. They were the type of comments that fueled Army careers. But these comments were not about an RA officer, they were about a lowly U.S. Army Reserve officer who had postponed his service as long as he could; a citizen soldier who never liked fighting; an ordinary man from a small town in the cow country of western New York.

A final combat efficiency report was written by MAJ Coleman and LTC Williams covering the time I was service platoon commander (Wingnut 6) after Diamond left. It contained similar ratings and comments, and I knew the door was open to me if I should decide to become a lifer. But that wasn't likely. I'd had enough real war to last a lifetime.

The summary of my Vietnam experience looked like this: total days in Vietnam = 365; total days of flying in combat = 166; decorations received included the Vietnam Cross of Gallantry, the Vietnam Service Medal, the Vietnam Campaign Medal, 23 Air Medals (including 1 for valor), 2 Army Commendation Medals (including 1 for valor), and 2 Bronze Stars (including 1 for valor). I also had received an altered perception of the world.

## Chapter 14 - Welcome Home?

I don't remember arriving at the Columbus Airport, or the ride home. In fact, I don't remember anything at all until a welcome home party at my wife's parents' house in Upper Arlington, Ohio. I didn't really want to be there, but it was the dutiful thing to do.

The guests at the party were mostly friends of my wife's family, people I knew before the war, some of whom were relatives and neighbors. They were cordial, but none could engage in the conversations I was used to having about things related to combat and the Army, conversation about life and death matters.

I was exasperated by the seeming unimportance of the subjects of their daily lives that they talked about. To my battle-trained ear it was all drivel and there was no place for me in this conversation. So I handled this frustration in the same way I handled the stress at Can Tho, I became inebriated. The only other thing I remember is either thinking, or maybe saying out loud, "You people don't know what the f--- you're talking about."

Not long after that party, I asked my father for permission to use his camp in the Adirondack Mountains near Old Forge, NY to get away for a few days. He agreed, and I called Tom and asked him to come along. A few days later Tom, my wife and I headed for the camp where we spent several quiet days decompressing and exploring the North Woods. It was a relief not to have to talk to anyone else.

Even before arriving home I felt like Dana Andrews playing the character, Fred, in the movie *The Best Years of Our Lives*. That was the story of three men trying to come to terms with their war experiences in a post World War II America. Those veterans came home as heroes after a brutal, but successful and seemingly necessary war. America was grateful for their service.

Just like Fred, I was trying to figure out how to come to terms with my war experiences. But it wasn't 1945 it was 1971, and we were in a very different America than the men who came home after WWII. Like them we came home from a brutal war, but unlike WWII, the Vietnam War had lost the support of Americans and had become an embarrassment. And while some people castigated veterans, I think most people just didn't know what to do with us; so they ignored us.

After more than a decade of bad news about American involvement in Southeast Asia, the country and I just wanted the whole subject to go away.

# Postscript - July 2013

## Getting Free

It's been a long time, more than two generations in fact, since the stories in this book took place. And during all those years, the subject of Vietnam and Southeast Asia, and what the U.S. did there, did not go away. It's even more present to me now than it was then, because the passage of time not only heals wounds and thereby erases obstacles, it gives perspective and understanding. It also provides a safe distance that allows connections to be made with uncomfortable memories, and uncomfortable truths.

When I began writing these stories in late 2010, I thought it would be a fairly short and easy project. After all, all I had to do was dump my memories on paper for my kids, and their kids, to read whenever and if ever they might be interested. But I was naive. The farther I got into my own history, the more the need to understand the context of those times became apparent. And that lead to reading and research that has consumed thousands of hours over the last thirty months.

Doing research in early 2012, the problem of UXO and landmine contamination in Southeast Asia, a problem created by the U.S. to a large degree, caught my attention. It was not the first time I'd heard about it; I'd even given a financial contribution to help with the problem a decade ago. But this time the problem seemed different, because I suspected America was being affected by this problem just as our brothers and sisters in Southeast Asia were. And as my knowledge and understanding grew, I began to connect the dots of cause and effect.

More specifically, there appears to be a relationship between what we did then and the discord we are experiencing in America now, and this situation may be an example of the expression "past is prologue." We seem to be experiencing an unintended legacy for the decisions made 40+ years ago, which causes a contemporary problem that erases the time and distance between America in 2013 and Vietnam in 1970.

In my research, I read a lot about how Vietnam fared after the country was unified in 1975. But I needed to experience the country for myself, so in February 2013 I returned to Vietnam with my old stick buddy, Tom Yost. Over a two week period we traveled from Hanoi to Soc Trang experiencing the country and the people first hand, but this time for peaceful purposes.

Six days into our trip, we were given a tour of the UXO and landmine removal operations run by the Mines Advisory Group in Quang Tri Province, near Gio Linh (*gee-oh lean*). It was both educational and eye-opening. The most sobering moment occurred in a farmer's field where an unexploded M-79 round was found laying in a fence row. While there, we also learned that the large depressions that pockmarked his rubber tree grove were bomb craters made by U.S. jets during the war.

I stood in that field with a light drizzle running down my raincoat as I listened to him recount what happened that day, many years ago. As he talked, I could hear the jets going over and feel the concussions of the bombs in the fields. I could hear M-16, M-60 and M-79 fire, and smell burned gun powder drifting through the air. And I could see this farmer, age nine at the time, huddled with his parents and sister in their A-frame bomb shelter, where a piece of shrapnel tore through his sister's scalp during the battle.

His description of that day was an unexpected flashback that was both a gift and an opportunity. With true humility I apologized for what we had done to him, his family and his country during the war. His quiet nod served as gracious acceptance of my apology, and I knew at that moment that I had paid my ransom, and was free.

**Standing in the Shadows**

We all understand that violations of the law have negative consequences. For example, if I run a stop sign I will pay a fine and maybe something more. But violations of principles can have negative consequences too.

I invite the reader to consider that the shadows of discord that America stands under today, the ones this book is named for and the ones I mentioned in the introduction are a metaphor for the consequences of our violation of the Principle of Good, which might also be thought of as the Law of Good.

As a practical example, let's look at the principle of mathematics. I know that 2 + 2 = 4, and no amount of argument or self-justification will permit 2 + 2 = 5 to work. To illustrate, if I add a deposit into my checkbook thinking 2 + 2 = 5, it may result in getting a letter from the bank indicating I have overdrawn my account and will be assessed a penalty, simply because I didn't obey the principle of math.

In order to correct this problem, two things have to occur. First, I have to pay the penalty, which is a $30 overdrawn fee in the above example. It's a $30 ransom, if you will, that returns me to good grace with the bank. And second, I have to learn how math works to avoid making the same mistake over and over until the bank closes my account.

The Principle (or Law) of Good works like the principle of math. If I violate it, I must pay the ransom and then learn how it works if I want to avoid future errors and more negative consequences. In such cases, ignorance of my obligation to the principle, in itself, will not remove me from the consequences of my mistakes.

America is standing in the shadows from Can Tho today. Those shadows are violations of the Law, or Principle, of Good, violations that we practiced in Vietnam two and a half generations ago that have now caught up to us. The government intrusion and ongoing terror we are now experiencing in America is the manifestation of "what goes around comes around." They are our equivalent of the unexploded ordnance and landmines that the U.S. left in Southeast Asia, showing up here in America.

So what can we do to change it? The same things we do for any violation of a law or principle: we pay the penalty (ransom) and change our ways.

**Paying the Penalty**

From the standpoint of fairness (another important principle) our invasion of Vietnam was an unjust act, and it doesn't matter how the politicians and statesmen justified it at the time. Once it was done and subsequent atrocities occurred, America was condemned by its own acts and the shadows of our own guilt eventually came home to us.

To be released from this condemnation of our own making, we must pay the required, multi-fold penalty and atone for our actions by (1) recognizing and accepting the mistake, (2) developing contrition for the injustice that resulted, (3) paying a ransom for the error, and (4) teaching ourselves and our children how to prevent such mistakes in the future. Only when these things are done can we return to harmony with the Principle of Goodness, and receive the benefits of that atonement. These actions are not possible without the prerequisites of humility, and obedience to the demands of that Principle. I know this because I have walked this path.

The first step, recognizing and accepting the mistake, requires brutal honesty with the self and with each other. An effective way for me to do that was to study the problems we created and then left in Southeast Asia when we pulled out. Reading David Halberstam's book *The Best and the Brightest* and researching the magnitude of the UXO and landmine problem at www.the-monitor.org were most helpful in this self-education.

The second step may be more difficult because almost no one wants to voluntarily experience the state of grief that being contrite requires. This could be especially true for Americans who were not involved directly in the war, or were not even alive at the time it occurred. But consider this metaphor: virtually none of the passengers on the Titanic had anything to do with the design of the ship, or the collision with the iceberg that sank her. Yet the lives of all passengers were altered by the sinking.

Humility and grief are often necessary to be able to acknowledge our failures, to develop contrition for their consequences, and to experience a desire to atone. Just reading Halberstam's book is a good start down this road. And in case a voluntary step into grief is intimidating, let me assure you that I have spent many moments of sorrowful measure experiencing the gift of tears in my study, pursuing the truth that this step reveals. (I would not trade this experience for anything.)

Step three, paying for the error, is both an individual act and an act of community, and it can happen in at least two ways. The first is to offer a sincere apology to the aggrieved party, in this case the citizens and the government of Vietnam. That can be done at no cost whatsoever by any individual who participated in the war, and it can also be done by the U.S. government which started the war. The second way requires money, to eradicate the UXO and landmines from all of Southeast Asia. The U.S. government provides some funding now, but at the rate clearance efforts are being funded it will take several hundred years to clear the land of this scourge.

Financial support of this cause by U.S. citizens and letters to our elected representatives and senators supporting a public apology to the people of Vietnam will go a long way toward lifting the shadow off our country. Other nations have issued public apologies for their mistakes. It's only appropriate that we do so for all the victims of our mistake in Southeast Asia.

Step four, teaching ourselves and our children how to prevent such mistakes, requires teaching the truth about the past. Even if that were done, however, it would not be enough to prevent a repeat. That will require an understanding of the nature of unintended consequences (of what goes around comes around) which must be taught to our children by us, because no one else will do it.

Most Americans I've talked with about the Vietnam War have no knowledge of the UXO and landmine problem. They are appalled to learn that this is our legacy in Southeast Asia and realize that, in effect, the war never ended there. And judging by their reactions, they are shocked to learn that America is guilty of such an ongoing atrocity and want something done about it.

**Changing Our Ways**

Upon his victory over the British fleet on Lake Erie in 1813, Commodore Oliver Hazard Perry proclaimed, "We have met the enemy and they are ours." In 1971, a sage comic-strip possum named Pogo stated, "We have met the enemy and he is us." In little more than 150 years America went from defending itself against foreign enemies on our own soil, to being its own worst enemy on others' soil, a situation that will continue until we learn to change our ways.

Americans today know that something is deeply wrong in our land, and many public polls report that a majority of Americans believe the country is on the wrong track. Regardless of what some may think, I know that Americans are not stupid, they are just infinitely patient as they wait for our leaders to rise to the public's level of understanding.

But we, the public, need to be more emphatic in demanding that our government learn how to act in the long term best interest of all the peoples of the world, and stop looking at other countries and peoples as targets of opportunity. Such a change requires our political and military leaders to develop a new reality, a reality that's closer to the heart.

| *Closer to the Heart* - Rush |
|---|

To get right on this issue, i.e. to accept the obvious consequences of our mistakes in Vietnam and Southeast Asia, I had to find and remove landmines, shadows, in my own soul. It's been hard work, but the reward has been immeasurable.

Finding and removing the landmines that are in our souls because of the war in Vietnam is the essential requirement for all Americans, and for the U.S. government, if we are to successfully dispel our shadows from Can Tho.

# GLOSSARY

0-0 (or zero-zero) - shorthand for a no-fly weather condition where the ceiling is at the ground and visibility is zero feet

1LT - the commissioned officer rank of first lieutenant

AAF - Army airfield

AC (or aircraft commander) - a designation that appoints a pilot to full command of an aircraft and/or an air mission

AFVN - Armed Forces Viet Nam; a shortened term referring to the radio station (AFVN radio) run by the U.S. military in Vietnam

AFB - air force base

AH-1G - a U.S. Army helicopter gunship, also known as a Cobra

AHC - assault helicopter company

AIT - advanced individual training; training in a soldier's military specialty that ran from several weeks to several months, depending on the specialty

AK-47 - a Russian-made rifle used by communist forces in Vietnam

AMOC - Aviation Maintenance Officers Course

AO - area of operations- the airspace and ground area where combat operations are proceeding

Ao Dai - a traditional tight-fitting silk tunic worn over pantaloons by Vietnamese women

Arc Light - the code name and general term for the use of B-52 bombers to support ground tactical operations

ASAP - as soon as possible

ARVN - Army of the Republic of Vietnam; also the name for South Vietnamese soldiers (pronounced *ar-vin*)

Ash & trash - a type of combat mission that delivers supplies or mail to soldiers in field locations by aircraft; also known as Direct Combat Support

AWOL - absent without leave (i.e. without permission)

Betel Nut - the seed of a palm tree that grows in the tropical Pacific that is often chewed wrapped in betel leaves; it's often used as a mild pain killer and stains the chewer's mouth and chin red

BOC - battalion operations center

BOQ - bachelor officers' quarters; buildings on U.S. Military bases for quartering commissioned officers

Bore sight - the alignment of weapons on an aircraft; on Huey's the bore sighting was usually aligned with the fuselage

Bounty Hunter - a gunship crew member in the 191st Assault Helicopter Company

Brass - high ranking military officers

C ration - a complete, cooked meal of canned/packaged food, usually consumed in field locations where meals cannot be cooked and served

C5A - a large U.S. Air Force cargo plane

C&C - Command and Control- a type of combat air mission

CA - combat assault; a tactical offensive maneuver involving helicopters placing infantry troops into landing zones

Chalk - a specific aircraft load, especially a group of airborne soldiers that deploy from a single aircraft

Charlie, or VC, or Viet Cong - guerrilla soldiers who fought against the United States in Vietnam representing the National Liberation Front

Chicken Plate - Kevlar body armor worn over the chest and abdomen

Claymore mine - a directional anti-personnel mine used by the U.S. military. Unlike a conventional land mine, the Claymore is command-detonated and directional, meaning it is fired by remote-control and shoots a pattern of metal balls into a kill zone, like a shotgun

CO - commanding officer

Collective - the control in a helicopter that sets the angle of attack for the main rotor blades, and thereby makes the aircraft climb or descend

Cowling - the metal covering over an aircraft engine

CPT - abbreviation for the rank of captain

CTA - Can Tho Airfield

CWO - Chief Warrant Officer; one of the advanced Warrant Officer ranks (CW2, CW3, and CW4)

Cyclic - the "stick" in a helicopter that controls the direction of flight by controlling the tilt of the rotor head

Dap, The - a ritual greeting among black American soldiers involving the knocking together of fists

DCS - direct combat support (see ash & trash)

DEROS, or Date of Estimated Return from Over Seas - the date U.S. soldiers were expected to return to the U.S. from duty in Vietnam

Dinky dau - Vietnamese term for "crazy"

DMZ, or Demilitarized Zone - a swath of land at approximately the north 17$^{th}$ parallel that divided north and south Vietnam, created by the 1954 Geneva Convention

"Don' mean nothin'" - a slang term for "everything is going to be alright"

Downwash - the downward airstream from helicopter rotor blades; the opposite of up-wash

Dustoff - see medevac

EM Club - enlisted men's club; a place where enlisted men find entertainment, usually on a military post or base

ENFANT System - an experimental, infrared detection system mounted on a Huey and used to detect enemy soldiers in Vietnam during darkness

Extraction - removing infantry troops from a landing zone

Fragging - from fragmentation grenade, refers to the act of murdering another member, or members of the military, particularly (a) member(s) of one's own command or fighting squad

G-Load - the force of gravity on an aircraft crew member

GCA - ground controlled approach; a method used in low visibility conditions whereby pilots receive verbal instructions from an air controller on the ground, who provides directions for landing

GI - Government Issuant; a short nickname for a soldier of enlisted rank

Green Deltas - pilots and crews assigned to the second platoon of the 191st Assault Helicopter Company

Guns - a shortened form of the word gunship; a nickname for the men who flew Huey gunships

H-3 - Hotel 3; the heliport at Tan Son Nhut Air Force Base near Saigon (now known as Ho Chi Minh City)

Hammerhead stall - a maneuver in which an airplane pulls up in a vertical climb until it almost stalls and then drops the nose in a wingover so that direction of flight is reversed

Hootch - an unsophisticated dwelling

Horn, The - the radio

Hornet's Nest - a dangerous place to be

Horse trading - trading goods or services in order to get goods or services that one is not authorized to have

Huey - UH-1; the model number of the primary utility helicopter used by U.S. forces in the Vietnam War

I Corps - the U.S. tactical area that encompassed the five northernmost provinces of South Vietnam

II Corps - the U.S. tactical area of 12 provinces that encompassed the Central Highlands of South Vietnam, north of Saigon

III Corps - the U.S. tactical area of 11 provinces, including the city of Saigon, north of the Mekong Delta in South Vietnam

IV Corps - the U.S. tactical area of 16 provinces south of Saigon that encompassed the Mekong Delta of South Vietnam

IFR - Instrument Flight Rules; flying under conditions where flying by visual references is obscured; flying by instruments only

Tactical IFR - flying under IFR conditions in a tactical (i.e. combat) environment

IP - instructor pilot

Insertion - placing infantry troops into a landing zone

JP-4 - fuel for jet engines

KBA - killed by air

Klick - a kilometer

LBJ - Lyndon Baines Johnson; or Long Binh Jail

Lightship - a Huey helicopter with a powerful xenon light mounted to the cargo floor, used to seek out enemy soldiers or positions at night and often used in combination with machine guns; also called a "super slick"

LZ - landing zone

M-16 - a U.S. military rifle that fires a 5.56 mm bullet

M-60 - a U.S. military machine gun that fires a 7.62 mm bullet

MACV - the acronym for Military Advisory Command Vietnam, which was the U.S. Army organization to which U.S. military advisors were assigned

MAJ - abbreviation for the rank of major

MARS - Military Auxiliary Radio System; a civilian auxiliary consisting primarily of licensed amateur radio operators who are interested in assisting the military with communications on a local, national, and international basis as an adjunct to normal communications

M.A.S.H. - Mobile Army Surgical Hospital; a United States Army medical unit serving as a fully functional hospital in a combat area of operations

Medevac or Dustoff - a helicopter crew whose mission was to rescue sick or injured troops

Mortar attack - an attack on ground forces using an indirect-fire, muzzle loading cannon, fired at high angles

MPs - military policemen

MPC - military payment certificate; a substitute for hard currency issued to U.S. troops in Vietnam

N1 topping check - a test flight diagnostic procedure to test the power of a jet engine

NCO - non-commissioned officer; the general title for enlisted men's ranks of E-5 (buck sergeant, or sergeant) through E-9 (sergeant major)

Number 1 - a slang term meaning "the best"

Number 10 - a slang term meaning "the worst"

NVA - North Vietnamese Army

O Club - officers club; a place where military officers find entertainment, usually on a military post or base

OCS - Officer Candidate School

OD - olive drab, the color of Army fatigues in Vietnam; or Officer of the Day, a commissioned officer who was responsible for the security of a military installation

OER - officer efficiency report

OH-6 - a U.S. Army helicopter used for low-level reconnaissance, also known as a Cayuse, or loach

OIC - Officer in Charge

PCS - permanent change of station

PE - periodic examination or inspection; a complete inspection of a Huey that was required to be performed every 100 hours of flight time. A PE required at least a few hours, and could last days if extensive maintenance work was found to be needed. Older aircraft usually required more time for PE

PFC - private first class, the second lowest enlisted men's rank in the U.S. Army

Piastre - the currency of French Indochina and early South Vietnam; eventually replaced with the South Vietnamese dong

PIC - Pilot in Charge

Prick 25 - a portable radio worn on a man's back

PSP - pierced steel planking; steel plates used to make temporary runways or roads

Puff - short for Puff the Magic Dragon, a high firepower U.S. C-46 fixed wing plane used in Vietnam

PX - post exchange; a store on a U.S. Army post that caters to military personnel

PZ - pickup zone, a place where troops or supplies were loaded onto helicopters

RA - regular Army, in contrast to the Army Reserve

Red-Xed - a designation that denotes an unsafe condition and usually prevents an aircraft from being flown

Revetment - a parking area for one or more aircraft that is protected by blast walls

R&R - rest and recuperation; a one-week time period allotted to each serviceman during his 12-month combat tour when he could leave Vietnam to visit Hawaii, Bangkok, Australia, Hong Kong, or other places

ROE - rules of engagement; a list of conditions that must be met before a suspected enemy could be engaged in combat

ROTC - Reserve Officer Training Corps

RPG - rocket-propelled grenade; a shoulder-fired, anti-tank weapon system that fires rockets equipped with an explosive warhead

RRF - Ready Reactionary Force; a military force organized to provide defense of a military position or facility

RU-6 (Beaver) - a U.S. Army fixed wing plane made in Canada by de Havilland Aircraft Corporation

S-1 - the U.S. Army personnel office for a battalion sized unit

Sampan - a narrow, flat-bottomed boat usually propelled by oars or a small engine, used in East and Southeast Asia

SFC - sergeant first class

Skid - one of a pair of metal bars used as the landing gear for a helicopter

Skycrane - nickname for a CH-54 heavy lift helicopter

Slick - a troop carrying Huey

SNAFU - an acronym for the U.S. Army term "situation normal, all fouled up"

Snake - nickname for an AH-1G Cobra gunship

SOI - Signal Operating Instructions; a type of military orders that control the communications within a command, containing technical details like radio frequencies

Sortie - a flight from one touchdown point to another

SP-4, SP-5, SP-6 - specialist 4th class, 5th class and 6th class: enlisted men's ranks in order of increasing rank and technical skill

Spooky - the Vietnamese counterpart of Puff

SSG - staff sergeant

Stinger - a metal rod about two feet long that was attached to the back end of the tail boom of a Huey; it protected the tail boom from damage that could result from a tail-low landing

There it is! - an expression used by GIs which meant "be cool" or "let it ride", because you can't change what can't be changed

Tracers - projectiles that are made with a small pyrotechnic charge in their base. Ignited by the burning powder, the pyrotechnic composition burns very brightly, making the projectile visible to the naked eye. This enables the shooter to follow the projectile trajectory to make aiming corrections

Translational Lift - a transitional state that occurs when a helicopter moves from a hover to forward flight; it usually occurs at about 15 knots of airspeed and can be felt by the crew inside the aircraft

UH-1C - a C model Huey used primarily as a gunship in Vietnam

UH-1M (M-model, or Mike model) gunship - a Huey equipped with an experimental, infrared heat detecting system known as an ENFANT System; only two helicopter units had M models in Vietnam, and the 191st was one of the two

UHF radio - ultra high frequency radio; a radio that operates in a frequency band of 300-3,000 megahertz

USARV - pronounced *use-r-vee*, it's an acronym for United States Army Republic of Vietnam

VC - Viet Cong, the communist guerrillas of South Vietnam; also known by the nickname "Charlie"

VFR - visual flight rules; a set of regulations under which a pilot operates an aircraft in weather conditions that are clear enough to see where the aircraft is going

VHF radio - a radio that operates in a frequency band of 30-300 megahertz

VNAF - Vietnamese Air Force

WO - Warrant Officer, a U.S. Army rank for technical specialists like pilots

World, The - slang term for The United States

XO - executive officer; the person who is second in command